TONY TILFORD

BIRDS OF BALI, SUMATRA AND JAVA

A PHOTOGRAPHIC GUIDE

H E L M

LONDON · OXFORD · NEW YORK · NEW DELHI · SYDNEY

HELM
Bloomsbury Publishing Plc
50 Bedford Square, London, WC1B 3DP, UK
29 Earlsfort Terrace, Dublin 2, Ireland

BLOOMSBURY, HELM and the Diana logo are trademarks
of Bloomsbury Publishing Plc

First published in the United Kingdom 2023

A catalogue record for this book is available from the British Library.
Library of Congress Cataloguing-in-Publication data has been applied for.

ISBN: PB: 978-1-4729-8687-0; ePub: 978-1-4729-8688-7;
ePDF: 978-1-3994-0926-1

2 4 6 8 10 9 7 5 3 1

Design by Susan McIntyre
Printed and bound in India by Replika Press Pvt. Ltd.

To find out more about our authors and books visit www.bloomsbury.com
and sign up for our newsletters.

CONTENTS

MAP OF THE REGION

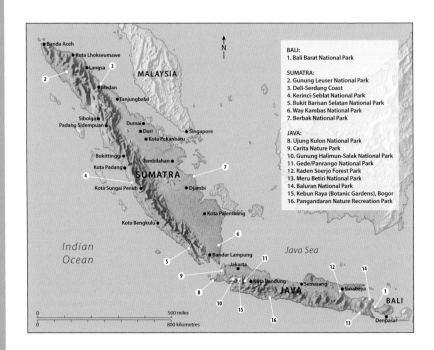

BALI:
1. Bali Barat National Park

SUMATRA:
2. Gunung Leuser National Park
3. Deli-Serdang Coast
4. Kerinci-Seblat National Park
5. Bukit Barisan Selatan National Park
6. Way Kambas National Park
7. Berbak National Park

JAVA:
8. Ujung Kulon National Park
9. Carita Nature Park
10. Gunung Halimun-Salak National Park
11. Gede/Panrango National Park
12. Raden Soerjo Forest Park
13. Meru Betiri National Park
14. Baluran National Park
15. Kebun Raya (Botanic Gardens), Bogor
16. Pangandaran Nature Recreation Park

INTRODUCTION

The islands of Bali, Sumatra and Java total some 613,000 square kilometres, almost 2½ times the area of the United Kingdom, and possess approximately 700 bird species. With such a rich avifauna, it is hardly surprising that birds are widely kept there for their beauty and song. Visitors, who now reach the region in ever-increasing numbers, cannot fail to notice that many houses have caged birds hanging from the eaves. Close inspection reveals that many of the species kept in captivity here are often the same as those once found in pet stores around the world. Fortunately, much of the international trading has now ceased, but local markets persist, taking their toll on local populations, with some being brought to the edge of extinction in spite of protective legislation.

So far, only the Javan Lapwing has become extinct in recent times, but several other species, most notably the Bali Myna, are hovering on the brink. The demise of the Bali Myna is well documented and is largely the result of illicit trapping, exacerbated by human disturbance and loss of habitat. The same dismal picture is repeated for many other species, including the Java Sparrow, Brahminy Kite and Black-winged Starling, all of which were, not many years ago, a common sight.

It is fortunate that the Indonesian authorities, in conjunction with national and international conservation organisations, have recognised the urgent need for protection, and vast areas have now been designated as national parks and reserves, with special protection for wildlife. Visitors should bear in mind that permits are often officially required to enter many of these reserves, and prior arrangements may be necessary. Even outside reserve areas the most threatened animals are protected, at least in theory.

Within the space limitations of this small book, it would be impossible to cover all the 700 or so bird species that have been recorded on the main islands of Bali, Sumatra and Java. The selection has therefore been based upon birds that are reasonably common throughout the region. A few less common but interesting and more spectacular species have been added to reveal the enormous biological diversity to be found there and to provide an understanding of its characteristic nature.

The 322 species treated include a reasonable representation from most of the families that may be encountered in the region. The reader will undoubtedly come across others, and these are covered in the more specialised (but unfortunately more cumbersome) books listed for further reading or on the internet.

BIRDWATCHING

For most of us, birdwatching is really putting a name to a bird, but there is so much more to be gained from more detailed study. Behaviour, life cycles and interaction with the environment are fascinating subjects which not only can give us enormous enjoyment but can also add much to our knowledge of the world around us.

Initially, however, identification must be the priority. We should consider all the clues available and, ideally, identify not just the species but also its sex, age and race. There are always many clues to go on, and by combining them we narrow down the possibilities until our problem is solved. We all accept the value of visual clues, but we tend to neglect the importance of sound until we become more experienced. For the seasoned birdwatcher, a sound recorder is just as important a part of his or her kit as are notebook, binoculars, telescope and field guide. That important call can often be identified later in the day.

Birdwatching aids

Whatever equipment you take, it should be well protected from the ravages of the weather. Rain and high humidity provide the worst conditions for cameras, binoculars and electronics. Any equipment should be kept dry and well aired and, if possible, stored with a bag of silica-gel desiccant to protect against fungal attack. It is astonishing how much damage fungus can cause if inadequate precautions have been taken.

Ideally, lightweight waterproof binoculars are recommended, particularly if you are going to venture into wet and humid areas, but for general use, the not too expensive kind are perfectly adequate for most observation. For the more serious birder, there are many other professional-style binoculars to choose from, as well as spotting scopes for more distant observation.

For sound recording, it is possible to use a smartphone for close-up sounds, but for better reproduction, use the compact, lightweight and reasonably priced Olympus recorders, such as the LS-P5. For more distant birdsong, an added external parabolic microphone is preferred to a directional 'gun' microphone if you want really good recordings.

As for cameras, they are not recommended for regular birdwatching trips, not only because of the possibly detrimental effect of the weather on equipment, but also because it is almost impossible to concentrate on both activities at once, particularly in a group situation.

It is fortunate that many of the good birdwatching areas are accessible by car. Also, there is always public transport and, except on Bali, at the end of

Common Green Magpie.

the journey the ubiquitous motorbike-taxis known as 'ojeks'. In the field, the energy-sapping heat and humidity are perhaps the biggest drawbacks. It should be unnecessary to advise here on hygiene and the need to carry sufficient bottled water. Keeping fit and healthy is extremely important, particularly in areas where medical facilities may be poor or non-existent. Do not forget to take essential medication with you, and always be prepared for stomach upsets. Remember, 'Bali Belly' is not confined to Bali. Insects can also be a problem at certain times, and insecticide sprays and creams could save you from some nasty bites. Suitable clothing is also essential, not only as protection against the elements but also as a disguise. Lightweight, dull-coloured clothes are necessary. Visitors are recommended to carry a light poncho with a hood, and in lowland forests long trousers are desirable for warding off leeches and stinging plants from making contact with your skin. Sturdy footwear is a must, although many people prefer heavy-duty sandals, particularly in wet conditions. Lightweight canvas boots are better than leather ones, which quickly become mouldy.

HOW TO USE THIS BOOK

Photographs depict the bird in a commonly seen form where possible, either male or female, and variations are briefly described in the text. The order and naming of the 322 species covered generally follows the International Ornithological Congress (IOC) taxonomy adopted by Avibase – The World Bird Database and the Cornell Laboratory of Ornithology current terminology. While other checklists exist, the process of revision based on more recent scientific knowledge will continue. At the time of writing, a concerted effort is taking place with the International Ornithologists' Union (IOU) forming the Working Group on Avian Checklists (WGAC) with a view to produce and maintain an open-access global checklist of bird species that will unify different taxonomies. It must therefore be expected that not only will common names vary, but so too will the scientific. Ultimately, the scientific names prove more reliable. The names used in this book follow the latest scientific data available as contained in the second edition of Eaton, J.A., van Balen, B., Brickle, N.W. and Rheindt, F.E. *Birds of the Indonesian Archipelago: Greater Sundas and Wallacea* (2021).

Each species description includes an English name and its scientific name and a formal Indonesian name where known. As a size guide, the overall length of the species from the tip of the bill to the end of the tail is provided. The remaining description follows no particular order but provides most of the clues necessary for a fairly reliable identification. Only for a few species has any attempt been made to describe calls and song, as it is felt that written descriptions lack meaning without the experience of hearing the actual sound. Nevertheless, the value of sound must not be underestimated. In many cases, it can be the only determining factor in successful species identification. Much of the description is necessarily brief and written in a semi-technical language adopted by most bird books, but for easy understanding many of the technical terms used by ornithologists are explained in the glossary on page 218.

A note on the photos

All images show the adult plumage, unless stated otherwise. Please refer to the key below for further information:

♂ – male	♀ – female
imm. – immature	juv. – juvenile
var. – variant	br. – breeding non br. – non-breeding

THE REGION

Java, Sumatra and Bali lie along the edge of the South-east Asian tectonic plate, where it rides over the Australian-Indian plate, generating intense geological activity. A line of volcanoes, some still active, others dormant, forms a spine along the island chain. On Sumatra these create the Barisan Mountains, stretching the entire length of the island, and falling precipitously into the depths of the Indian Ocean in the west; in the east it drops more gently, giving rise to large swampy areas before reaching the Sunda Sea. On Java, the mountains arise in isolation from alluvial plains and, as in east Sumatra, slope gently northwards to the Sunda Sea.

At one time Bali was connected by a land-bridge to Java, but today it is separated by some 3km of turbulent waters, renowned for their fierce currents. Not surprisingly, the fauna and flora of Bali show close similarities to those of eastern Java. Sumatra, however, shows closer affinities with Borneo than with Java.

The fragmentation of the landmasses combined with geological and climatic activity has resulted in the formation of many habitats, which in turn support a rich diversity of wildlife. This includes many endemic species: those confined to a single region or smaller area and found nowhere else.

Java and Bali are among the most densely human-populated regions on Earth, and the destruction of the remaining wildlife habitats is a matter of serious concern. Only on remote mountain slopes can one find significant areas of natural forest. A few patches of lowland forest survive in national parks and nature reserves. Sumatra, with only a tenth of the population density, is marginally better off, but the growth of agriculture, plantation and forestry is responsible for its increasing forest destruction.

The avifauna

We might expect that these landmasses, which were at various times joined to mainland Asia, would have similar flora and fauna. As the various islands were not all separated simultaneously, however, and because subsequent local climatic conditions have varied, the composition of the avifauna differs significantly from one island to another. Perhaps the most dramatic changes have been caused more recently, by man and his destruction of the lowland forests. There are ample records dating back as far as 1657, and it is quite clear from the reports of the early 19th-century explorers who founded Indonesian ornithology that birds were far more plentiful then.

When and where to find birds

As the region lies close to the equator, the climate is hot and humid. The monsoon weather patterns of the region are characterised by wet and dry seasons, which differ little in temperature. Even so, the birds show distinct breeding cycles. During the wet season, from October to March, large numbers of wintering migrants arrive from the north, with a smaller number of Australian species appearing during the dry season.

Many waterbirds arrive at the end of the wet season, nesting in the safety of isolated trees in a flooded landscape. This is also the peak of the insect breeding season, when abundant food is available for insectivorous species to rear their young. Frugivorous species delay breeding a little longer, until the trees and bushes are bearing ripe fruit.

Good birding areas, with a rich diversity of habitats, abound throughout the region. Clearly, it is not possible to list them all in a book of this size, but to guide the birdwatcher to some of the more productive areas, the major reserves and national parks are described below. These are shown on the accompanying map at the front of the book.

Chesnut-bellied Partridge.

KEY BIRDWATCHING SITES

BALI

Bali Barat National Park

Situated at the western tip of Bali, Bali Barat is a mix of dry and moist forests with coastal scrub, savanna and mangroves. It is the only place to find native wild, endemic Bali Mynas. Many other interesting species reside there, among them Banded Pitta, Green Junglefowl, Pink-headed Fruit Dove, Black-backed Fruit Dove, Shiny Whistling Thrush, and Sunda Bush Warbler.

Nusa Penida Island

Nusa Penida Island, while strictly speaking in Bali province, is just east of the Wallace Line. Here you may see Lemon-bellied White-eye and Wallace's Elegant Pitta.

SUMATRA

Gunung Leuser National Park

This is a huge area of rainforest and mountains with a small coastal and lowland extension. It is an ideal place to explore the birds of northern Sumatra, such as Wreathed, Black, Oriental (or Asian) Pied, Helmeted and Rhinoceros Hornbills, Great Argus, Crestless Fireback, Crested Partridge, Brown, Yellow-crowned and Gold-whiskered Barbets, White-rumped Shama, Black-and-yellow, Green and

Dusky Broadbills, Jay Shrike and Red-bearded Bee-eater.

Kerinci-Seblat National Park

The park includes the dominating peak of Mt Kerinci and the Kerinci valley wetlands. The best place to see Sumatran montane birds, as well as several of the endemics such as Schneider's Pitta, Sumatran Peacock Pheasant and Sumatran Cochoa. Also recorded there are Salvadori's Pheasant, Silver-eared Mesia, Chestnut-capped Laughingthrush, White-throated Fantail, Sunda Bush Warbler and Golden Babbler.

Way Kambas National Park

Situated in the south-east of Sumatra, Way Kambas includes remnants of

Wallace's Elegant Pitta.

lowland rainforest and coastal and swamp forest. Many species have been recorded there, including White-winged Duck, Storm's Stork, Stork-billed Kingfisher, Lesser Adjutant, Crested Fireback, Crested Partridge, Great Argus, Hill Myna, Black-bellied Malkoha, Scarlet-rumped Trogon, Gould's, Sunda and Large Frogmouths, Black-thighed Falconet and Orange-breasted and Cinnamon-headed Green Pigeons.

Berbak National Park

This reserve is on the east coast and includes coastal forest and peat swamp, as well as mangroves. It is a good area for Büttikofer's Babbler, Wallace's Hawk Eagle and Milky and Storm's Storks.

Bukit Barisan Selatan National Park

Situated at the southern tip of Sumatra, at the end of the Barisan Mountains, the park reaches from the sea to the top of Gunung Pulung and includes all forest types and some very wild areas. It is relatively unvisited but could yield some interesting surprises. Among the rarer birds in the area are Sumatran Treepie, Helmeted Hornbill, Red-billed Partridge, Lesser Adjutant and Milky and Storm's Storks.

JAVA

Ujung Kulon National Park

At the extreme west tip of Java, Ujung Kulon boasts many rare species in a great variety of habitats, from lowland rainforest and evergreen forest to coastal scrub, mangroves and open grazing areas. As many as 24 endemic or threatened species are to be found there, including Javan Coucal, White-breasted Babbler, Javan Sunbird, Javan Hawk Eagle, Blue-throated Bee-eater and Green Peafowl.

Javan Hawk Eagle.

Javan Barred Owlet.

Green Peafowl.

Gede/Pangrango National Park

This West Javan national park consists mainly of luxuriant evergreen submontane forests, but with mossy elfin forest and alpine meadows at the peak of Gunung Pangrango. It is a wonderful place to see many of the Javan endemics, such as Javan Tesia, Javan Cochoa, Javan Hawk Eagle, Chestnut-bellied Partridge, Javan Scops Owl, Javan White-eye, Dusky (Horsfield's) Woodcock, Pygmy Tit, Mountain Serin, Volcano Swiftlet and Blue-tailed Trogon.

Meru Betiri National Park

Meru Betiri covers large areas of moist primary and secondary forests, mangroves and a rather rugged coastline with some beautiful beaches. It is one of the less explored areas where new discoveries are likely to be made. Among the rarer birds recorded are Wreathed Hornbill, Banded Woodpecker, Violet Cuckoo, Pin-tailed Parrotfinch, Black-crested Bulbul, Grey-cheeked Green Pigeon, Black-naped Fruit Dove, Javan Owlet, Waterfall (or Giant) Swiftlet, Black-banded Barbet and Crescent-chested Babbler.

Baluran National Park

Situated at the north-east tip of Java, Baluran National Park covers the environs of the dormant volcano of Gunung Baluran. It is one of the easier places for birdwatching, being more open and easily accessible, and is a good place to see Green Peafowl and Green and Red Junglefowls. White-bellied Woodpecker, Oriental Pied Hornbill, Spotted Wood Owl, Banded Pitta, Lesser Adjutant and Grey-cheeked Tit Babbler are all reported from the area.

Kebun Raya (Botanic Gardens), Bogor

Being easily accessible, Bogor's Botanic Gardens become crowded with local visitors at the weekend. A weekday visit, however, can be very fruitful in the comparative tranquility. The enchanting settings of Kebun Raya, covering 1km^2, have been established for around two centuries and provide very ready opportunities to see many town and woodland birds. There is a roost of Black-crowned Night Herons,

and it is difficult to miss Black-naped Oriole, Sooty-headed Bulbul and Oriental Magpie Robin. Other species likely to be seen are Black-naped Fruit Dove, Collared Kingfisher, Blue-eared Kingfisher, Grey-cheeked Green Pigeon, Horsfield's Babbler, Ornate Sunbird, Purple-throated Sunbird, Hill Blue Flycatcher and Yellow-throated Hanging Parrot.

Pangandaran Nature Recreation Park

Just west of Cilacap on the south coast of Central Java is the peninsula of Pangandaran, with its nature recreation park. Consisting mainly of dry forest, it is a good area for Green Junglefowl and Javan Banded Pitta. Other species present are Oriental Pied Hornbill, White-rumped Shama, Scaly-crowned Babbler and Black-backed Kingfisher. To the east lie the mangroves of Segara

Ornate Sunbird.

Milky Stork.

Anakan, where many waterbirds can be seen. Among the large population of herons, egrets and terns, Lesser Adjutant and Milky Stork may be seen. Javan Coucal and Stork-billed Kingfisher have also been recorded there.

VISITING NATIONAL PARKS

Fees to enter national parks can be paid at the entrances. In most Indonesian national parks, trail networks, experienced bird guides, and accommodation close to the best birding sites are limited.

Javan Pied Starling.

Wandering Whistling Duck *Dendrocygna arcuata* 40–45cm
Indonesian name: Belibis Kembang

Plumage is generally a deep chestnut-brown on the back, tail and breast but with white undertail-coverts and rump, and a line of black-edged white feathers showing below the folded wing. The head and neck are a paler brown, and there is a dark brown elongated cap running down the back of the neck. The legs are blackish-brown and bill black. This bird emits a series of high-pitched whistling and twittering calls in flight.

Where to see: Rivers, freshwater lakes, wet marshland, coastal wetlands and mangrove throughout the region.

Cotton Pygmy-goose *Nettapus coromandelianus* 30–35cm
Indonesian name: Trutu Kapas

A small black-and-white duck, the male being predominantly white, with iridescent black plumage on the crown, back, wings and tail and a black neck-band. The female's plumage is more subdued, being buffish-white where males are white, and brown where males are black; she has a brown eye-stripe and lacks the neck-band. In flight, males show a white wing patch.

Where to see: Throughout Java, Bali and northern coastal wetlands of Sumatra at freshwater marshland and lakes, flooded paddyfields and grassland. Scarce. Unlike most wildfowl, this species regularly perches in trees and normally nests in tree holes.

Garganey *Spatula querquedula* 37–41cm

Indonesian name: Itik Jurai

Males seen in the region are usually in eclipse plumage looking much like females. They have brown-barred feathers and a contrasting head pattern, with pale eyebrow, dark eye-stripe, and plain whitish throat. Neck and breast are reddish-brown flecked white. It has a green upperwing speculum with a white border above and below. The elongated scapulars are striped grey, dark green, black and white, and the upperwing-coverts are light blue-grey. The flanks are finely patterned grey and black and the belly is white.

The bill is blackish-grey, the eyes reddish-brown and legs and feet greyish.

Where to see: Occurs throughout the region, but scarce; on open marshland, freshwater lakes, rice fields, coastal salt-marshes, mudflats and inshore waters.

Pacific Black Duck *Anas superciliosa* 47–60cm

Indonesian name: Itik Alis

The name 'black duck' is a misnomer, as the bird's body plumage looks black only in flight when contrasting with the conspicuous white underwing plumage.

It is actually dark brown, with the head striped black and white. The speculum is iridescent green and purple, the legs yellow-brown and the bill grey.

Where to see: Occurs throughout the region but is less frequent on Sumatra. It is a surface-feeding duck which dabbles in shallow water. Confined mainly to mountain lakes of East Java and Bali, but often found feeding on marshes and grassland.

Roulroul *Rollulus rouloul* 25cm
Indonesian name: Puyuh Sengayan

The very distinctive male sports an ostentatious tufted, spiky red crest above a white crown patch. His overall plumage is blackish-blue with a purple sheen, shaded green towards the tail; the wings are a dark red-brown. Females lack the crest and have green body plumage and a grey head; the wings are dark chestnut. They have red legs and bare red skin around the eye. The bill is black, with males showing more red at the base.

Where to see: Lowland peat bogs and hill forests of Sumatra, where it congregates in family groups, foraging for seeds, insects and fallen fruit.

Chestnut-bellied Partridge *Arborophila javanica* 27cm
Indonesian name: Puyuh-gonggong Jawa

Endemic to Java, each subspecies has a different head pattern of reddish-buff with black markings and a black collar. The breast is grey varying through chestnut to a white vent. The flanks are chestnut, the back and tail grey-brown and barred black, and wings brown with black barring and small white spots. The legs are red and the bill blackish.

Where to see: Occurs in clearings in montane forests on Java, usually in pairs but occasionally in small groups, foraging amongst undergrowth.

Great Argus *Argusianus argus* male 160–200cm, female 70–75cm

Indonesian name: Kuau Raja

One of the most spectacular of the region's pheasants, particularly the males with their extraordinarily elongated secondary and tail feathers used in their mating display. With tail raised high and wings spread wide, the beautiful patterning of green 'eye' marks (ocelli) is spectacular. Plumage is otherwise mainly chestnut-brown, broken and patterned by spots and flecks of buff and black. The female is generally darker, has shorter wings and tail, and lacks the male's eye spots. Both sexes have a short, dark brown crest and bare blue skin on the head and neck.

Where to see: Undisturbed primary rainforests of Sumatra. Often a single male accompanied by several females.

Green Peafowl *Pavo muticus* males: 180–250cm females: 100–110cm

Indonesian name: Merak Hijau

This spectacular pheasant is predominantly iridescent green, males having conspicuous ocellated tail feathers and a vertical crest on the head. Females are similar but duller with shorter legs and tail and paler underparts. Juveniles appear similar but have a few white feathers on the face and neck. During the breeding season their mewing and loud calls are readily heard, and the male's striking display with tail feathers fanned is unmistakable.

Where to see: A few local relict populations exist on Java, with Baluran National Park being the last stronghold.

Blue-breasted Quail *Synoicus chinensis* 12–15cm
Indonesian name: Puyuh Batu

The male of this species is distinctive, but females are easily confused with Small and Barred Buttonquails. The yellow feet are the main distinguishing feature. The upperparts are dark brown with lighter streaks, the female being paler than the male. Males have a distinct black-and-white bib pattern, with the breast, sides of the head and flanks a rich grey-blue; the belly and undertail-coverts are bright chestnut. The female's underparts are brown streaked buff, with darker brown barring across the chest and a whitish throat patch and buffish eye-stripe. The newly hatched young are tiny, almost like bumblebees.

Where to see: Occurs sparsely throughout the region's lowland grasslands, savannas and marshes, sometimes in cultivated areas.

Red Junglefowl *Gallus gallus* males: 65–78cm, females: 41–46cm
Indonesian name: Ayam-hutan Merah

Males are recognised by their serrated red comb, face and wattles, their long bronze hackles and long, dark green, arched tail feathers. They have a chestnut to golden mantle and blackish-green breast and primary coverts. Females are much duller; various shades of brown with black streaking on the neck. The bill is buff and the legs slate-grey. Separating wild birds from domestic or feral ones is not easy, but purely wild birds have tail and wing and collar colour all black in males, brown in females, body feathers all black in males, mottled brown in females, dark grey legs and bright orange hackles.

Where to see: Forests and scrublands of north Sumatra, Java and Bali, sometimes near human habitation.

Green Junglefowl *Gallus varius* males: 65–75cm, females: 40–46cm
Indonesian name: Ayam-hutan Hijau

This large blackish-green fowl is similar to the Red Junglefowl but with an unserrated purplish-red comb. Males have red wattles and bare facial skin around the eye, a glossy green mantle and nape and iridescent green hackles edged yellow, with uppertail- and wing-coverts orange and yellow; the rest of the plumage is black. Females and young juveniles have a less showy plumage with brown upperparts sparsely mottled with buff, buffish underparts mottled with black and a white throat.

Where to see: Endemic to Java and Bali. Occurs close to the coast, also inland forest, preferring more open, grassy areas and often associating with grazing animals.

Sunda Collared Dove *Streptopelia bitorquata* 30–33cm
Indonesian name: Dederuk Jawa

Not be confused with the similar Eurasian Collared Dove (*Streptopelia decaocto*). Generally appears as a pinkish-fawn bird with a broad black half-collar fringed with white. The crown and nape are pale bluish-grey tinged pink and the forehead very pale grey. The upperparts and wing-coverts are slightly darker pink. Underparts are pale pink fading to white at the vent and undertail-coverts. It has a pinkish-brown bill, pink legs and a brownish-orange iris.

Where to see: Occurs mainly in the lowlands of Bali and Java, often in open areas but also close to human habitation. Frequently seen foraging on the ground.

Eastern Spotted Dove *Spilopelia chinensis* 30cm

Indonesian name: Tekukur Biasa

This very familiar dove has a generally pinkish-brown plumage with slightly darker flight feathers and dark-mottled back. On each side of the neck it has an obvious white-spotted black patch. The outer tail feathers are broadly tipped white, this being conspicuous in flight. It has red feet and a black bill.

Where to see: Occurs throughout the region, from city gardens to rural cultivation, open country and forest edges in lowland areas, often foraging on the ground.

Barred Cuckoo Dove *Macropygia unchall* 37–41cm

Indonesian name: Uncal Loreng

Distinguished from other ground doves by its largish size and heavily barred upperparts, this dove is rather dull-coloured; the upperparts are brown, heavily barred with black. The underparts are paler with a buffish throat and pinker with finer black barring diminishing towards the vent. The head and nape are greyish-brown, with males having a green iridescence. The tail is blackish-brown, strongly barred reddish-brown. It has short, black bill, red feet and yellow or pale brown iris.

Where to see: Submontane forest throughout the region. Uncommon.

Parzudaki's Cuckoo Dove *Macropygia emiliana* 30–37cm

Indonesian name: Uncal Buau

Much smaller than the Barred Cuckoo Dove, with more of a uniform chestnut-brown head. The male's upperparts are a uniform brown with a light purplish iridescence on the upper breast and neck. The underparts are a slightly paler chestnut-brown with light black barring on the breast. Females are similar in colour but have black barring on the back and the wing-coverts and slight barring on the upper breast.

Where to see: Found in the hill forests of Java and Bali, where it prefers forest edges and clearings with low scrub. Uncommon.

Little Cuckoo Dove *Macropygia ruficeps* 29cm

Indonesian name: Uncal Kouran

Smaller than the other brown cuckoo doves, it is separated by its buff breast and by the dark brown subterminal bar on its outer tail feathers. The upperparts also are lightly barred black and the upper breast is often heavily barred, especially on the female. Males have green and purple iridescence on the nape. They have red legs and a brown-tipped black bill. Females are duller with less iridescence.

Where to see: Common in hill and montane forests throughout the region. Often in large flocks, at forest edges and feeding in grassland and ripening rice fields.

Asian Emerald Dove *Chalcophaps indica* 23–27cm

Indonesian name: Delimukan Zamrud

This attractive ground dove has iridescent emerald-green wing-coverts and mantle, with a dark grey lower back crossed by two conspicuous white bars. The flight feathers and short tail are black. The underparts, neck and sides of the head are dark pink. Males have a white forehead, with the crown and nape tinged grey. Females are duller. The bill is bright red, the legs and feet dull pink and eyes dark brown.

Where to see: Occurs throughout the region, preferring the seclusion of thick forests, woodland edge and mangroves.

Doves and pigeons

Barred Dove *Geopelia maugeus* 22–26cm

Indonesian name: Perkutut Loreng

The Barred Dove is often confused with the Zebra Dove being similar in appearance but with a distinctly different song. Also, it has stronger black and white barring stretching across the breast and onto the flanks, a white belly and chestnut underwing-coverts, bare yellow skin around the eye and a pale yellow iris.

Where to see: Occurs from Bali eastwards, inhabiting dry open scrub and grassland, cultivated land and woodland edges in lowland areas and even cities.

Zebra Dove *Geopelia striata* 20cm
Indonesian name: Perkutut Jawa

Known as the Zebra Dove for its finely barred pale brown and black plumage from neck to tail. The back of the head is pale brown blending into a buff face, crown, neck and chin. The long, tapered tail is brown and black with white-tipped outer feathers; very evident on take-off. Commonly known locally as the Peaceful Dove, it should not be confused with its close relative, the Peaceful Dove (*Geopelia placida*) of Australasia.

Where to see: Occurs sparsely throughout the region. Seen around gardens, cultivation and open grassland and scrub. This is a very common cagebird on Bali and Java, where it holds a special significance in cultural mythology.

Grey-cheeked Green Pigeon *Treron griseicauda* 25cm

Indonesian name: Punai Penganten

Appears as a yellowish-green pigeon with a grey head and either a green back in the female or a maroon back in the male. Both sexes have a bare green-coloured skin patch around the eye and a grey forehead. The blackish flight feathers are edged with yellow. The central tail feathers are green with darker outertail feathers showing a greyish subterminal bar. The undertail-coverts are chestnut in the male but paler in the female. The male's bill is yellow with a green cere, unlike the green bill of the female. The legs are red.

Where to see: Endemic to Bali and Java. Found high in the trees in lowland forests, sometimes with other pigeons in fruiting trees; sometimes close to human habitation.

Pink-necked Green Pigeon *Treron vernans* 27cm

Indonesian name: Punai Gading

This small green pigeon is distinguished from similar pigeons by a grey tail with black band and pale grey tip. The male has a blue-grey head merging through pink to orange on the lower breast; the abdomen is green and yellow, the back and wings green. The female is green, lacking the male's brighter colours, but identified by tail pattern and keeping company with the distinctive male. Its calls are a 'cooing' whistle and when feeding, a series of crow-like rasps.

Where to see: Common throughout the region in lowland and coastal forest and mangrove, and in more open countryside. It becomes very conspicuous when trees are fruiting.

Black-backed Fruit Dove *Ptilinopus cinctus* 38–44cm

Indonesian name: Walik Putih

A very distinctive black-and-white fruit pigeon with yellowish lower belly and vent. The head and neck are greyish-white blending to pure white on the upper breast. A wide black bar separates the upper breast from the grey belly. The upperparts are very dark greenish-black blending into a dark grey-green tail with a grey terminal bar. The feet are red and the bill grey. Its flight is rapid, with fast, noisy wingbeats.

Where to see: Found on Bali in moist subtropical lowland and montane forest, often feeding from fruiting trees.

Pink-headed Fruit Dove *Ptilinopus porphyreus* 29cm
Indonesian name: Walik Kepala-ungu

The male's head, neck and throat are a delightful purplish-pink, fringed with a contrasting white band across the breast, then a greenish-black band followed by a grey expanse of belly leading into yellow undertail-coverts. The upperparts are green, feet pink, bill pale green and the iris orange. Females have a dull pink forehead and face, the remainder of the head, neck and throat being green; and more obscure black and white chest bands.

Where to see: Occurs throughout the region's montane forests. Shy and inconspicuous. Occasionally it congregates in large numbers in fruiting fig trees

Black-naped Fruit Dove *Ptilinopus melanospilus* 27cm
Indonesian name: Walik Kembang

This species has predominantly green upperparts, tail and lower breast with a yellow vent and red undertail-coverts. The male has a pale grey head with a conspicuous black nape and yellow throat. It has red feet and a greenish-yellow bill. Females and juveniles have the head all green, with juveniles having obvious yellow feather edges. Usually detected by its resonant *ow-wook-wook, ow-wook-wook* call.

Where to see: Locally common in the lowland and hill forests of Java and Bali and irregularly on Sumatra. Seldom seen except when it congregates in flocks at fruiting trees.

Green Imperial Pigeon *Ducula aenea* 42cm

Indonesian name: Pergam Hijau

Head, neck and underparts are soft pinkish-grey, with chestnut undertail-coverts. The upperparts are green with a bronze iridescence. It has a dull red base to the blue-grey bill, and dark red feet.

Where to see: This is the commonest pigeon of the region's lowland rainforest. Inhabits riverine forest and coastal mangroves, where it prefers to live high in the treetops, feeding on figs, nutmeg and other small fruits.

Mountain Imperial Pigeon *Ducula badia* 43–50cm
Indonesian name: Pergam Gunung

Distinguished by its large size and dark maroon-brown upperparts with light vinous-grey head, neck and underparts. The throat and crown are whitish in adults.

The tail is grey with a broad lighter grey terminal band and undertail-coverts are buffish-white. Legs and feet are red, the bill dark red and the grey or white iris is surrounded by dark red orbital skin. Juveniles are duller with less warm colouring. Not to be confused with the Dark-backed Imperial Pigeon (*D. lacernulata*) on Java, which has a well-defined grey crown and sides to the head.

Where to see: Occurs in the montane forests of Sumatra and West Java.

Pied Imperial Pigeon *Ducula bicolor* 35–42cm
Indonesian name: Pergam Laut

This is a striking bird, mainly creamy to ivory-white, with black flight feathers and tip of tail. The bill and feet are blue-grey. Immatures are greyer. Hunting has decimated the population of this beautiful pigeon, particularly on Java and Bali.

Where to see: More often in coastal forests and mangrove, it sometimes congregates in substantial breeding colonies in areas where fruiting figs are readily available.

Javan Coucal *Centropus nigrorufus* 46cm

Indonesian name: Bubut Jawa

Javan Coucal has a black back, inner secondaries and wing-coverts with chestnut-brown outer primaries and

primary coverts. Otherwise all black with a distinct purplish sheen, even its tail. The bill and legs are black and the iris red. In size it is midway between the Greater Coucal and Lesser Coucal and can be separated from them by the black back and some wing feathers.

Where to see: Endemic to Java. Uncommon. Usual habitat is mangrove forest, swampland and dry shrubland and Alang-alang grassland around the coast.

Greater Coucal *Centropus sinensis* 47–53cm

Indonesian name: Bubut Besar

This large, bluish-black, crow-like bird with chestnut-brown wings and back, and long, wide, black tail is most likely to be seen lumbering about in small trees and bushes, hiding away in thick undergrowth. The bill and feet are black and the iris brown. Females are similar to males but juveniles are lightly barred buff and brown overall. Its eerie, far-reaching, low-pitched call starts with a series of *poop, poop, poop* sounds, increasing in tempo then slowing again, but occasionally a series of just four deep monotonous *poop, poop, poop, poop* sounds are made.

Where to see: Throughout the region, where it prefers forest edges, grassy banks, secondary scrub and even village gardens and cultivated land.

Lesser Coucal *Centropus bengalensis* 31–35cm

Indonesian name: Bubut Alang-alang

One of the smaller coucals, having a very long hind toe claw. It has dark plumage, mainly black with light rufous wings and a long black tail. In breeding plumage the head and upper back are shiny with dark feather shafts, but at other times it is duller and the feather shafts appear whitish. This feature extends to the wing-coverts, which have an appearance of white-streaked brown feathers. Juveniles have red-brown barred upperparts, with feather shafts appearing buffish-white, the underparts are buff and undertail-coverts barred black.

Where to see: Occurs throughout the region, mainly in open lowland marshy areas among Alang-alang grasses and low tree cover.

Chestnut-breasted Malkoha *Phaenicophaeus curvirostris* 42–50cm

Indonesian name: Kadalan Birah

Easily recognised by its large, curved bill; the upper mandible is pale yellow and lower mandible brownish-red or black. There is a patch of bare red skin around the eye. In the male, the iris is pale blue, but in the female it is yellow. The head is grey with a thin white supercilium and underparts and rump chestnut. Wings are dark green to dark blue and the feet dark grey to black. Juveniles have a yellow bill with partially black lower mandible.

Where to see: Inhabits dense undergrowth of damp primary and mangrove lowland forest throughout the region, usually seen feeding on small vertebrates, insects, small crustaceans and sometimes figs.

Red-billed Malkoha *Phaenicophaeus javanicus* 42–44cm

Indonesian name: Kadalan Kembang

Separated from other malkohas of the region by its large size and strong red bill. The upperparts are mid-grey but tinged with a pale bluish-green iridescence. The underparts are chestnut with a dark grey band across the upper breast, and the grey tail feathers are tipped white. The eye is surrounded by a patch of bare blue skin and the legs are grey. Juveniles have a black tip to a paler red bill. Can be located by its whining call.

Where to see: Occurs in the lowlands and lower hills of Java and Sumatra in secondary scrub and at evergreen forest edges; usually in pairs or small groups, foraging in Alang-alang grassland.

Horsfield's Bronze Cuckoo *Chrysococcyx basalis* 17cm

Indonesian name: Kedasi Australia

A small cuckoo recognised by its green and bronze iridescent shading on its otherwise partially brown-barred upperparts and its white eyebrow with a brown eye-stripe. The pale buff underparts are lightly streaked brown at the neck, becoming stronger on the flanks and undertail-coverts but whitish at the centre of the belly. The feet and eye-ring are grey, the thinnish bill is black and the iris red-brown. Juveniles are generally duller with much less barring and a paler grey-brown iris.

Where to see: Occurs as a migrant in March and April on Bali and Java and occasionally in south Sumatra, usually in coastal, drier open woodlands.

Banded Bay Cuckoo *Cacomantis sonnerati* 22–23cm

Indonesian name: Wiwik Lurik

Adults occurring on Java and Bali have bright rufous-brown, barred darker upperparts and buffish-white, finely barred brown underparts. They have a broad buff-white supercilium above a brownish eye-stripe with a whitish line below. The iris is grey-brown. The bill is black and feet are grey. Juveniles differ in having white-tipped feathers on the crown and mantle. Birds on Sumatra are darker and appear more iridescent bronze on the upperparts. They show more yellow on the legs and the iris is much paler.

Where to see: Occurs throughout the region. Found at lowland forest edges. Uncommon.

Plaintive Cuckoo *Cacomantis merulinus* 18–23cm

Indonesian name: Wiwik Kelabu

Adults are pale grey over the head, neck and upper breast contrasting with the dark grey wings and upperparts. The grey on the breast blends into a lighter grey and then rufous-buff, becoming buff at the vent. The tail feathers are grey, tipped white. The red-brown iris is encircled by a narrow whitish eye-ring. The bill is black and the legs yellow-brown.

Where to see: Occurs throughout the region's lowlands, especially in sparse woodland and sometimes in towns and villages.

♂ Sumatran

♀ Hepatic morph

Sunda Brush Cuckoo *Cacomantis sepulcralis* 21–28cm

Indonesian name: Wiwik Uncuing

Adults are grey above with a grey head. The wings are grey-brown, slightly glossed green. The tail feathers are black with white tips and what appear as white horizontal V-shaped patches along their outer edges. The chin is grey above a plain chestnut upper breast becoming paler towards the vent. The iris is brown and the eye-ring bright yellow, the feet are yellow-buff and the bill is black, occasionally yellow below. Females show slight grey barring on the lower breast. Juveniles are duller with brown-barred plumage.

Where to see: Occurs throughout the region, usually at forest edges and in isolated mangroves.

Drongo Cuckoo *Surniculus lugubris* 23–25cm

Indonesian name: Kedasi-hitam

This mainly black cuckoo has a square-cut forked tail like a drongo. Its upperparts are shiny black with brownish-black underparts, apart from white bands on the undertail-coverts and all-white thighs. The bill is black, the legs are grey and the iris is grey-brown. Females are duller with yellow irises. Juveniles are similar but spotted white on the head, breast and wings. Its loud, clear calls are an ascending scale of warbled notes.

Where to see: Occurs in lowland forests throughout the region.

Large Hawk-cuckoo *Hierococcyx sparverioides* 38–40cm
Indonesian name: Kangkok Besar

The adult of this, the largest of the hawk-cuckoos, has brown upperparts and a brown-barred tail with blackish subterminal band and fine white tip. The head is grey with a prominent yellow eye-ring and an orange iris. There is no white nape like other hawk-cuckoos. The upper breast and belly are white-buff heavily streaked and barred brown, with the upper breast tinged russet. The lower belly and vent are buff-white. The upper mandible is blackish and lower mandible black to yellow at the base. Legs are orange-yellow.

Where to see: Occurs throughout the region, mainly in lowland woods and scrub, occasionally close to cultivation.

Common Cuckoo *Cuculus canorus* 32–34cm
Indonesian name: Kangkok Erasia

Adults are dark grey above with paler grey nape, crown, chin and upper breast. The breast and belly are white, thinly barred black and becoming more russet at the vent. The blackish tail feathers are spotted and tipped white. The iris is orange to light brown encircled by a yellow eye-ring. Legs and feet are yellow. Bill black above and yellow below, becoming darker at the base. Females generally show a warmer colouring to the breast but also occur in a reddish-brown (hepatic) form, in which the rump and uppertail-coverts are rufous and upperparts show black and brown barring.

Where to see: Overwinters in the region. Can occur anywhere.

Large Frogmouth *Batrachostomus auritus* 39–43cm
Indonesian name: Paruh-kodok Besar

This is the largest of the Asian frogmouths, quite varied in colouration with both sexes very similar. The upperparts are warm brown to buff-brown with pale-barred scapulars and dark-bordered buffish spots on the wing-coverts. The outer wing and tail are brown with paler bars. The underparts are brown spotted buffish with a paler belly. The bill is dark buff becoming brown at the tip. The legs are yellow and the iris is brown. Females are less strongly marked and a bit duller. Juveniles are lighter coloured.

Where to see: Occurs in moist lowland primary rainforests and second growth on Sumatra. Its numbers are declining.

Javan Frogmouth *Batrachostomus javensis* 20–23cm
Indonesian name: Paruh-kodok Jawa

Upperparts are generally a mixture of brown, buff, grey and white with black spotting; underparts are brown with buff, cinnamon and white speckles, bolder on the breast. The paler belly and flanks are lightly barred brown. The scapulars are marked with buffish-white patches, but the wings and tail lack white spots. It has a brown bill, brown legs and yellow iris. Females have warmer and brighter colouration but show less white. It is generally insectivorous, gleaning food from foliage, on the ground and sometimes in flight.

Where to see: Typically occurs in dense undergrowth in swampy areas of the tropical moist evergreen lowland forest of Java. Usually spends night high in treetops. Roosts closer to the ground during the day.

Large-tailed Nightjar *Caprimulgus macrurus* 25–29cm

Indonesian name: Cabak Maling

This large grey-brown nightjar is concealed by well-camouflaged plumage, its tail reaching far beyond its wings. In flight, the broad white tips to the two outer pairs of tail feathers and the distinctive white patch covering the centres of the four outer primary wing feathers of the male aid identification. The upperparts are grey-brown streaked dark brown and black, and there is a vague buff nuchal collar. A large white throat area and white moustachial streak contrast against the brown underparts, which are barred with buff at the neck to brown at the belly. The bill and legs are black and iris is brown. The female has buff-spotted wings and tail.

At dusk, it utters a slow and repeated *choink-choink-choink* call.

Where to see: Occurs throughout the region's mangroves, forest edges and scrubland, usually spending the daytime in shade on the ground.

Savanna Nightjar *Caprimulgus affinis* 20–26cm

Indonesian name: Cabak Kota

One of the smallest nightjars in the region, this species has a monotonous repetitive call, *jweep*, heard around dawn and dusk, often from birds in flight. The well-disguised, uniform brown plumage has a buffish-white nuchal collar and small white patch on either side of the neck. The male has white outer tail feathers and a sizeable white spot on its four outer primary wing feathers. Bill is dark grey and feet are reddish-brown. The female has no white in the tail and dirty buff wing spots.

Where to see: Occurs throughout the region. Common in dry open areas around the coastal lowlands, but is also found in big cities exploiting flying insects attracted by the lights.

Edible-nest Swiftlet *Aerodramus fuciphagus* 11–13cm

Indonesian name: Walet Sarang-putih

Difficult to separate from other small swifts in flight. Plumage is overall blackish-brown, appearing greyer on the throat, underparts and rump. Sumatran birds are often darker.

Where to see: They naturally inhabit the region's dark caves, where they use echolocation for navigation. Today, this is the swiftlet that has become big business in Indonesia, where they are encouraged to nest in disused buildings so their nests can be harvested to sell for bird's nest soup.

House Swift *Apus nipalensis* 14–15cm

Indonesian name: Kapinis Rumah

This medium-sized swift can be recognised by its broad white rump, pale throat and slightly notched tail, unlike the similar but larger Pacific Swift (*Apus pacificus*), which has a distinct fork to the tail. Otherwise, it has blackish plumage, black bill and brown feet.

Where to see: Usually seen in large groups hunting insects over open countryside. It prefers coastal habitats and the lower hills and is frequently found in towns and villages, where it nests under the eaves of houses. It also uses overhanging cliff faces and cave entrances as nesting sites.

Whiskered Treeswift *Hemiprocne comata* 15–17cm
Indonesian name: Tepekong Rangkang

Smaller than the Grey-rumped Treeswift (*Hemiprocne longipennis*), which also occurs in the region. Has a long, slightly forked tail and rudimentary crest. Males have a mostly dark greenish-bronze body, shiny blue-black wings, head and throat offset by long white supercilium and moustachial stripes separated by dark chestnut ear-coverts. The belly and tertials are whitish. Females are much the same but with dark blue-green ear-coverts. Juveniles are duller and browner.

♂

Where to see: Occurs on Sumatra, mainly in lowland primary and secondary forest or woodland with tall trees.

Common Moorhen *Gallinula chloropus* 30–35cm
Indonesian name: Mandar Batu

This very adaptable aquatic bird is almost entirely dark grey to black, with a broken white stripe along its flanks. It also displays white patches under the tail, particularly when dashing for cover with its tail raised. It has a characteristic red frontal shield and bill, with a yellow tip to the latter. The legs are greenish-yellow.

Where to see: Frequents the region's freshwater pools, lakes, rivers and rice fields. It is a good surface swimmer and will dive and even run over the water's surface.

Purple Swamphen *Porphyrio porphyrio* 40–48cm

Indonesian name: Mandar Besar

Easily recognised, large, blue-black rail with a substantial bright red bill and frontal shield, the upperparts having a purple-blue sheen on the mantle and sides of the neck. The upperwing-coverts are black shaded with green and its undertail-coverts are white. Its long, ungainly pink legs with huge spreading toes are ideal for trampling through and over aquatic vegetation. It habitually flicks its tail.

Where to see: Occurs throughout the region. A rather quiet and secretive bird, keeping to the edges of undisturbed reedbeds, marshland and occasionally paddyfields.

White-breasted Waterhen *Amaurornis phoenicurus* 28–33cm

Indonesian name: Kareo Padi

A conspicuous rail with a white face, foreneck and breast along with rufous flanks and lower belly well-demarcated from the dark grey-green upperparts. The bill is greyish-green with a red base, and the legs and feet are yellow. It is very vocal at dawn and dusk, with a cacophony of weird squawks, grunts and croaking sounds.

Where to see: Occurs throughout the region's wetlands. It is a wandering feeder, frequently out in the open but never far from dense cover. Very agile, it runs over floating vegetation and clambers through bushes, searching for insects.

Red-legged Crake *Rallina fasciata* 22–25cm

Indonesian name: Tikusan Ceruling

The red-brown mantle and brighter red-brown head along with the red legs are diagnostic of this rail. The remaining upperparts are grey-brown. The underparts and forewings are barred black and white. It has a green bill. Female similar but has a more pinkish head with narrower barring on the lower parts of the body. Immatures have less obvious markings and paler underparts and brownish legs.

Where to see: Occurs throughout the region's damp lowland forests and disturbed forest edges, and sometimes nearby cultivation.

Ruddy-breasted Crake *Zapornia fusca* 21–23cm

Indonesian name: Tikusan Merah

Easily confused with other crakes of the region, it is the only one that has both head and breast bright chestnut, with a white chin and blackish-brown underparts finely barred white. It has a green bill and red legs. Upperparts are plain russet-brown.

Where to see: A relatively common resident found throughout the region's lowlands. As a partially nocturnal and shy bird, it prefers the seclusion of paddyfields and reedy habitat and often bushland adjacent to water.

Beach Thick-knee *Esacus magnirostris* 51–57cm

Indonesian name: Wili-wili Besar

This stocky shorebird is recognised by its substantial black bill, which is yellow at the base, and its face, heavily patterned with black and white stripes. The crown is dark chestnut, the upperparts brown, and neck and upper belly buff lightly streaked fawn. The belly to undertail-coverts are buff. The long and thick, rather gangly legs are yellow. It has a large yellow staring eye.

Where to see: Occurs as a resident shorebird throughout the region. Seldom found inland, preferring sandy, muddy or rocky coastal beaches, where it finds crabs and small crustaceans.

Black-winged Stilt *Himantopus himantopus* 35–39cm

Indonesian name: Gagang-bayang Timur

This tall and rather elegant shorebird has long pink legs that trail behind in flight. It has a long white neck and small head with a slender, straight, black bill. The upperparts are black with a slight green iridescence and underparts white with some black-grey patches. The white head often shows varying amounts of black. Females and juveniles have duller upperparts.

Where to see: Occurs mainly in north Sumatra and West Java, preferring particularly shallow wetland, swamps and rice fields and occasionally coastal saltmarshes.

imm.

♂ var.

Plovers

Grey Plover *Pluvialis squatarola* 27–30cm

Indonesian name: Cerek Besar

In winter plumage, this rather attractive wader has upperparts mottled grey and brown and underparts pale grey to white. It has a faint pale supercilium, a black bill, dark grey legs and a large black eye. In flight, it shows a pale wing-bar and rump and, below, a white underwing with a large black patch near the body. These features easily separate it from the Pacific Golden Plover, which is slightly smaller and has a shorter bill.

Where to see: Occurs as a winter visitor to the region. Found in small groups on tidal mudflats feeding on marine invertebrates and often roosting on isolated shingle spits.

Pacific Golden Plover *Pluvialis fulva* 23–26cm

Indonesian name: Cerek Kernyut

A typical plover in shape, with an upright stance, a robust body, thin neck and largish head with a short, strong bill. In the region it is usually in non-breeding plumage, having lost the distinctive black face and underparts it had on its breeding grounds. Can be identified by its speckled gold, brown and buff upperparts, and buff breast, face and supercilium. The bill is black and the legs grey.

Where to see: This winter migrant stops off around the coasts of the Sundas, often congregating in flocks. It is common on mudflats and open grassland near the coast.

Greater Sand Plover   *Anarhynchus leschenaultii* 22–25cm

Indonesian name: Cerek-pasir Besar

9**Plovers**

This brown, white and grey plover joins
other waders, especially the smaller
Tibetan Plover on mudflats and sandy
estuarine beaches. It is most easily
separated by its more substantial,
slightly bulbous bill and from all
other wintering plovers likely to be
encountered as they show a breast-
band or a collar and have shorter legs.
Only early arrivals may show remnants
of the breeding plumage and have a
russet breast bar.

Where to see: Occurs on the mudflats
and beaches of Bali and Java as a
winter visitor.

non-br.

br.

Kentish Plover *Anarhynchus alexandrinus* 15–18cm
Indonesian name: Cerek Tilil

In breeding plumage, males have a black line passing below the eye extending over the ear-coverts and a black peak to the rufous crown and nape. The underparts, chin, cheeks, forehead, eye-stripe and collar are all white. There are black areas either side of the breast, and the upperparts are sandy-brown. The bill and legs appear black. Females and non-breeding males are more nondescript, with the male's black areas becoming dark brown.

Where to see: Occurs throughout the region, mainly on flat coastal beaches, lakes and waterways, and seldom on rocky shores.

Tibetan Plover *Anarhynchus atrifrons* 18–21cm
Indonesian name: Cerek-pasir Tibet

Easily confused with the larger, longer-legged Greater Sand Plover, especially when seen alone. It also has a more rounded head and shorter and thinner bill. Adults have a full black mask under a chestnut crown and a white throat above a broad rufous breast-band extending onto the flanks. The legs are grey. Non-breeding birds are much duller brown and lack the black and rufous markings. Juveniles are similar.

Where to see: Occurs as a winter visitor throughout the region's flat coastal beaches and mudflats.

Greater Painted-snipe *Rostratula benghalensis* 23–28cm

Indonesian name: Berkik-kembang Besar

A plump wader with a slightly decurved dirty-yellow bill (black-tipped in the female) and prominent narrow white band from the shoulder to the breast. Strong sexual dimorphism is reversed from normal. Larger female shows brighter and stronger characteristics, with longer blackish-green wings and dark chestnut upperparts, including the lower head, neck and breast. A white streak extends backwards from a white eye-ring. Legs are yellow. The male is duller, showing buff streaks on the wings and back.

Where to see: Occurs throughout the region on coastal and freshwater marshes and grasslands.

Pheasant-tailed Jacana *Hydrophasianus chirurgus*
40–58cm in breeding plumage, otherwise 30–31cm

Indonesian name: Burunq-sepatu Teratai

Breeding adults show a long, black tail and black underparts along with a brown back and scapulars that contrast with the lower white wing patch. The outer primaries are black. The front of the head and neck are white, separated from the yellow nape and hindneck by a thin black line. The legs are pale grey. When not breeding the tail becomes much shorter and upperparts are browner. The lower breast turns to white while retaining some of the blackish plumage at the lower neck. Immatures are more sombre-plumaged, similar to non-breeders, lacking the yellow on the neck and having a russet-brown crown.

Where to see: A rather rare migrant to the region's wetlands.

non-br.

Eurasian Whimbrel *Numenius phaeopus* 40–46cm
Indonesian name: Gajahan Penggala

A large, long-legged wader with a long neck, comparatively short, decurved bill and grey legs. Its plumage is heavily mottled brown, with a prominent black crown-stripe above a buff eyebrow and dark eye-stripe. In flight, it shows a white underwing, rump and lower back. Normally quiet, it will often give a trilling call when alarmed.

Where to see: A common winter visitor occurring throughout the region. Usually congregates on tidal mudflats and estuaries with flocks of other waders and occasionally on coastal marshes and rough grassland. Non-breeding birds are often present throughout the year.

Eurasian Curlew *Numenius arquata* 50–60cm
Indonesian name: Gajahan Erasia

This curlew is extremely well camouflaged. Plumage is variable but mainly light greyish-brown with dark streaking. It has a white belly and rump, and the legs are grey-blue. The long, brown, decurved bill is longest in the female. In flight curlews show a prominent white wedge shape on the rump and white below the wings. The call from which it gets its name is the recognisable *curloo*, often emitted in flight.

Where to see: Occurs as a winter visitor to the region's wetlands, grasslands and beaches.

Bar-tailed Godwit *Limosa lapponica* 37–40cm

Indonesian name: Biru-laut Ekor-blorok

Both this species and the similar Black-tailed Godwit (*Limosa limosa*) can be present at the same time. They are largish, long-legged waders with straight and long, upcurved bills and greyish-brown upperparts; Bar-tailed is more heavily mottled and has an obvious barred tail and slightly heavier grey streaking on the whitish breast.

non-br.

Where to see: An uncommon winter migrant to the region, generally found with other waders on tidal mudflats, estuaries, open expanses of sandy beach and saltmarsh. Perhaps more common on the east coast of Sumatra.

Ruddy Turnstone *Arenaria interpres* 21–26cm

Indonesian name: Trinil Pembalik-batu

This thickset little wader breeds in the Arctic. It has a strong bill and a rather blotchy appearance, being mottled black and white on the head, neck and breast. Upperparts are chestnut with buff lining and black patches. The underparts are white and the legs red. When overwintering in the region the plumage becomes duller.

Where to see: Overwinters throughout the region and is frequently observed on sandy and rocky beaches dislodging small rocks and stones to locate prey.

♂ br.

Red Knot *Calidris canutus* 23–25cm

Indonesian name: Kedidi Merah

More likely to be seen in its winter plumage; light grey all over with thick black legs and a short, straight, tapered bill. Other distinguishing features are its typical hunched feeding posture and to some extent its comparatively short neck, small head and eyes.

non-br.

Where to see: Overwinters throughout the region, regularly occurring at traditional coastal stopover sites, such as sandy and muddy inlets and estuaries, where it can easily locate small molluscs, crustaceans and other marine invertebrates.

Red-necked Stint *Calidris ruficollis* 13–16cm

Indonesian name: Kedidi Leher-merah

This small wader has upperparts marked with dark and paler shades of grey-brown, with the upperwings appearing greyer. The crown, neck and breast are grey with darker streaking and the face and underparts totally white. It has a short black bill and short, dark legs sometimes blotched orange and without webbing between the toes. Similar to the Little Stint (*Calidris minuta*), which occurs only very infrequently in the region.

br.

Where to see: A common migrant occurring throughout the region, usually seen in large wader flocks on estuaries, beaches and coastal mudflats, and on freshwater inland pools and rice fields.

Sanderling *Calidris alba* 20cm

Indonesian name: Kedidi Putih

In non-breeding plumage, it has pale grey upperparts and white underparts, with black bill and feet. A black shoulder patch helps to distinguish it from other waders. It shows black primaries and a prominent white wing-bar in flight.

non-br.

Where to see: Occurs throughout the region as an uncommon non-breeding migrant, found almost entirely on sandy beaches, usually in small groups following the retreating waves as they feed on marine invertebrates thrown up by the waves.

Common Snipe *Gallinago gallinago* 25–27cm

Indonesian name: Berkik Ekor-kipas

Easily confused with Pintail (*Gallinago stenura*) and Swinhoe's Snipes (*Gallinago megala*), and very difficult to separate. It has a long, black-tipped, buff-coloured bill and buff-white underparts heavily spotted warm brown on the flanks and belly. The upperparts are brown with buff feather edges creating a striped pattern from the mantle to the rump. It has a dark brown eye-stripe and broad buff supercilium. Throat and neck are also buff. The legs are yellow. Fast, erratic flight with very pointed wings.

Where to see: Occurs sporadically on Java and Bali in open freshwater marshland, estuaries, saltmarshes and rice fields.

Common Sandpiper *Actitis hypoleucos* 19–21cm

Indonesian name: Trinil Pantai

This sandpiper has brown upperparts and white underparts, with a brown patch extending from the neck to the flanks. In flight, its white and brown

barred outer tail feathers are evident, as is the white wing-bar. This is a very busy feeder with a constant tail-bobbing action. When disturbed it often flies a short distance and continues feeding, seemingly unperturbed.

Where to see: A common migrant throughout the region. Usually alone, feeding on coastal mudflats and along muddy riverbanks, and occasionally visiting marshland and rice fields upstream.

Grey-tailed Tattler *Tringa brevipes* 23–27cm

Indonesian name: Trinil Ekor-kelabu

It has uniform fawn-grey upperparts and is slightly lighter grey on the crown, neck and breast. The belly, flanks and undertail-coverts are white. The chin and neck are white to grey. The short legs are yellow and the bill black with a yellowish base. Females are slightly bigger than males with juveniles appearing spotted white on the upperparts.

Where to see: Typical of most migrant waders, it regularly appears on intertidal mudflats and shingly shorelines but also on rocks where beds of seagrass are present. A common migrant more prevalent in the east of the region.

Common Redshank *Tringa totanus* 27–29cm

Indonesian name: Trinil Kaki-merah

This long, red-legged wader has grey-brown flecked black upperparts from crown to rump. The throat, breast and belly are paler with more visible black flecking. The tail is barred grey-black and white, the bill is red at the base and tipped black, and the black iris is encircled by a white eye-ring.

Where to see: A common migrant to the region, occurring widely on estuaries, saltmarshes and inland freshwater wetlands.

Barred Buttonquail *Turnix suscitator* 16–17cm

Indonesian name: Gemak Loreng

The male, easily confused with the female Blue-breasted Quail (page 21), is mottled brown with breast and flanks barred with black. The main distinguishing features are the bright yellow legs and bill of the quail. Females are larger, having darker (almost black) chin, throat and crown but with more white on the head. During the breeding season, females can be heard making a brief repeated purring call at night.

Where to see: Occurs commonly throughout the region, preferring lowland open grass and scrubby habitat.

Oriental Pratincole *Glareola maldivarum* 23–25cm

Indonesian name: Terik Asia

Recognised by its mainly fawn-brown plumage with a white rump and dull yellow bib edged with a vague black necklace. The upperparts are a little darker brown with paler streaks on the crown, the breast is dark fawn blending to white at the vent and the flight feathers and forked tail are black. The legs are grey-black, the large iris is black and bill is black at the tip and dull red at the base.

Where to see: A common winter migrant, occurring throughout the region. Usually seen in open areas near water: tidal mudflats, estuaries, rice fields and pastureland.

Black-headed Gull *Larus ridibundus* 37–43cm

Indonesian name: Camar Kepala-hitam

The underparts and neck are white and upperparts are grey, including the back, upperwing-coverts, secondaries and inner primaries. The outer primaries are white tipped black, the secondaries are grey tipped white and the tail is white. Apart from the white nape, the head is blackish-brown contrasting with a broken white eye-ring. The bill is dark red and the legs pale red. Non-breeding birds are white shaded blackish-brown on the nape and ear-coverts.

Where to see: Occurs as a rare winter visitor, generally close to the coast but also lowlands inland and on lakes, rice fields and pastures.

Sooty Tern *Onychoprion fuscatus* 36–45cm
Indonesian name: Dara-laut Sayap-hitam

The upperparts and upperwings of adults in breeding plumage are black from the crown and nape and over the back with the outer tail being edged white. The forehead is white to just above the eye as are the underparts. The wing has dark primaries. In flight it shows a deeply forked tail while the white underwing-coverts contrast with the dark primaries. The iris is brown and the legs and bill are black. The non-breeding plumage is paler or browner and immatures are similar blackish-brown apart for the white vent, undertail and paler wing-coverts.

Where to see: Occurs around the region's coastline but seldom close to shore.

imm.

Bridled Tern *Onychoprion anaethetus* 37cm

Indonesian name: Dara-laut Batu

This is a medium-sized tern with a dark grey-brown back, tail and wings, white underparts and a long forked tail. The crown, nape and eye-stripe are black, leaving a thin white forehead leading into a short white supercilium but extending beyond the eye. Not to be confused with the Sooty Tern. The leading edge of the wings is white, as are the outer tail feathers.

Where to see: Occurs around the region's coasts, usually feeding on small fish or floating invertebrates taken from the surface of the sea, but seen ashore more during the summer months.

Lesser Crested Tern *Thalasseus bengalensis* 37–43cm

Indonesian name: Dara-laut Benggala

This species looks very similar to the Greater Crested Tern (*Thalasseus bergii*), but is slightly smaller and has a distinctive orange bill. In breeding plumage, the black cap extends forward to meet the bill. After breeding, the forehead and forecrown become white, leaving a black crest. Juveniles appear more like non-breeding adults, but with greyish wings and mottled brown upperparts.

Where to see: Occurs as a common winter visitor to Java and Bali, but less so to Sumatra. More often observed along sandy and muddy shorelines, but occurs often far out to sea.

Black-naped Tern *Sterna sumatrana* 30–33cm

Indonesian name: Dara-laut Tengkuk-hitam

Apart from the conspicuous black band through the eye and around the nape and its very pale grey upperparts, this small tern appears very white. In flight, its very long, forked tail is obvious. The legs are black, and the black bill is tipped yellow. Juveniles, however, have grey-brown mottled plumage over the upperparts and on the crown, with black mottling on the nape; the bill tip is also brownish and the tail unforked.

Where to see: This quite common resident occurs throughout the region, normally breeding on offshore rocky islands

White-tailed Tropicbird *Phaethon lepturus* 40cm

Indonesian name: Buntut-sate Putih

Adults are usually white, some with a yellowish tinge, and have long white tail-streamers. They are marked with black on the wing-tips and have a black bar on the upperwing. Juveniles have no tail-streamers, and their upperparts are barred black.

Where to see: Although resident in parts of the region, visitors also arrive from other areas. Individuals occurring around Sumatra, and recognisable by their golden plumage, originate from Christmas Island. Some breed on Nusa Penida Island, Bali, and along Java's south coast.

Asian Woollyneck *Ciconia episcopus* 76–91cm

Indonesian name: Bangau Sandang-lawe

This large, predominantly black, stork has a fluffily feathered white neck. The undertail and lower belly are also white, as are the forehead and a narrow eyebrow. It has a patch of mid-grey facial skin. The feet are a dirty red, while the bill is blackish, tinged red, with a red tip. Immatures differ from adults in having the black plumage tinged brown.

Where to see: Occurs as a migrant throughout the region, but uncommon on Java and Bali. May be seen feeding in rice fields and pastureland in the company of other storks, and often roosting with them in tall trees.

Storm's Stork *Ciconia stormi* 75–90cm

Indonesian name: Bangau Storm

A very rare stork, similar in appearance to Woolly-necked Stork. It has predominantly black upperparts, a black breast, white neck with black patterning at the side, and a black crown. The tail and belly are white. During the breeding season, the bright pink facial skin patch and the yellow eye-ring are very obvious. The slightly upturned bill is reddish and the legs are pink.

Where to see: Found only at freshwater pools in lowland forest in east Sumatra.

Lesser Adjutant *Leptoptilos javanicus* 120–130cm
Indonesian name: Bangau Tongtong

This strange-looking creature is not a pretty bird. Its massive horn-coloured bill on a virtually bald head and patches of bare orangey skin on the head, neck and upper breast give it the appearance of being deformed and sick. The back, wings and tail are dark grey and the underparts and upper back are white. The bill and legs are greyish. Juveniles have brownish-grey upperparts and dirty white underparts.

Where to see: Occurs throughout the region's coastal wetlands, mangroves, flooded grasslands and rice fields. Rare because of hunting and habitat loss on Java and Bali. It remains locally common in lowland areas of south Sumatra.

Milky Stork *Mycteria cinerea* 95–100cm
Indonesian name: Bangau Bluwok

This large stork usually occurs in single-species groups but occasionally in the company of other storks and herons. It is primarily white but has black flight feathers and tail. It has a prominent bare patch of facial skin which varies in colour from pink to red, a long, decurved, olive-yellow bill and greyish legs. Juveniles are a dirty pale brown, with the rump white and the flight feathers black.

Where to see: A relatively rare species but found throughout the region. It occurs in mangroves and mudflats of west Sumatra, the north-west coast of Java near Jakarta, and Brantas Delta, near Surabaya.

Christmas Frigatebird *Fregata andrewsi*
Length 90–100cm, wingspan 210–230cm

Indonesian name: Cikalang Christmas

This Critically Endangered seabird has very dark, brownish-black plumage, with long wings and a deeply forked tail. The male has a white patch extending between the belly and the vent, while in the female the patch is more extensive, extending forward to encompass the whole belly and onto the underwings. Males have a prominent bright red inflatable gular sac used in display, and grey bill and legs. Females are slightly bigger and have pink legs, bill and gular pouch. Immatures have variable head colouration, from dirty yellow to white.

Where to see: This vagrant occurs around coasts of the region and originates from Christmas Island.

Oriental Darter *Anhinga melanogaster* 87–95cm
Indonesian name: Pecuk-ular Asia

Very similar to cormorants apart from the long, snaking neck, small head and yellow-brown, dagger-like bill. It is mainly black, with white-streaked plumed coverts on the wing and back. It swims very low in the water with only its head and neck showing, so it often becomes waterlogged and has great difficulty in taking to the air. It spends a lot of time alone on low, exposed perches, often with its wings outstretched to dry, but it roosts communally.

Where to see: Occurs throughout the region, preferring large, clean freshwater lakes and slow, forested rivers.

Little Cormorant *Microcarbo niger* 51–56cm
Indonesian name: Pecuk-padi Kecil

The smallest of the cormorants in the region. The plumage is generally greenish-black, with a few tiny white feathers on the neck and head. The bill is yellowish-brown, with a black tip and purple near the base in adults. After breeding, the plumage darkens and loses the white flecks, except for the chin and throat, which become white. Immatures are dull brown.

Where to see: Occurs throughout the region, frequenting mangroves, marshland, lakes and estuaries, often in small flocks swimming low in the water and diving for fish. Once a common resident of the coasts, mangroves and lowland waterways of Java, but becoming rarer.

Little Pied Cormorant *Microcarbo melanoleucos* 58–65cm

Indonesian name: Pecuk-padi Belang

This little cormorant has a long tail and short, thick bill. In the region it loses its black-and-white pied breeding plumage for more blackish-brown and white. The underparts, neck, throat and sides to the face are all white and the upperparts blackish-brown. The legs are black, iris dark grey, and bill orangey-yellow, tinged green.

Where to see: Non-breeding birds occur on Java and Bali, favouring freshwater habitat, rivers, lakes and swamps. Also appears at coastal estuaries and mangroves.

Little Black Cormorant *Phalacrocorax sulcirostris* 55–65cm

Indonesian name: Pecuk-padi Hitam

A bird with uniform black plumage and a purplish-green iridescence, except for mottling on the wing-coverts. In the breeding season, it acquires a small patch of white feathers behind the blue-green eye. It has a small greyish gular pouch and bare skin around the eye. The bill is grey-blue and the feet black.

Where to see: Occurs as a resident on Java and Bali and is spreading along the east Sumatran coast. Mainly a bird of inland lakes, fish ponds and reservoirs, it is occasionally seen on coastal estuaries, where it roosts on rocks and trees. It is uncommon, but a colony exists at Pulau Rambut on the north-west Javan coast.

Australian Pelican *Pelecanus conspicillatus* 152–185cm

Indonesian name: Undan Kacamata

Similar in size to the Great White Pelican (*Pelecanus onocrotalus*) and identifiable by its black tail and upperwing-coverts, pinkish gular pouch stretching the full length of the very large pinkish-purple bill, and lack of bare facial skin patch. The primaries and secondaries are black, legs and feet grey-blue. It flies with laboured wingbeats and catches fish by plunging into the water.

Where to see: It breeds in Australia, migrating north and west as far as Bali and East Java. Occasionally found singly but more often in small flocks on estuarine sand bars and mudflats.

Cinnamon Bittern *Ixobrychus cinnamomeus* 40–41cm

Indonesian name: Bambangan Merah

Adult males have bright cinnamon-brown upperparts and orange-buff underparts and are streaked black on the centre of the belly and along the upper flanks. The sides of the long neck are streaked dirty white. Females and immatures appear darker, with a black cap, and immatures have a more mottled appearance. One of the smaller, more timid bitterns of the region. When disturbed, it rises from cover with a croaking alarm call and flies off with slow, powerful wingbeats.

Where to see: Occurs throughout the region, living in grassland, freshwater swamps and rice fields. Common.

imm.

Grey Heron *Ardea cinerea* 85–100cm

Indonesian name: Cangak Abu

A predominantly grey bird with paler underparts, it has a prominent black streak on the head extending into a small nuchal crest, black flight feathers and a line of black streaks on the front of its white neck. Legs and bill are yellowish. It appears smaller than the Purple Heron and lacks the reddish-brown colouration of that species.

Where to see: Found mostly in the region's lowlands, frequenting wet areas such as rice fields, lakes, mangroves and swamps, hunting fish, frogs and crabs. It is a colonial nester, often seen in tall trees in the proximity of mangroves.

Herons, egrets and bitterns

Purple Heron *Ardea purpurea* 75–90cm

Indonesian name: Cangak Merah

Similar in height to the Grey Heron, this more elegant, slimmer bird has a less hunched stance. Easily distinguished from other herons of its size by its darker underparts and its black-streaked, russet-coloured neck and upper breast. The back and wings are greyish-purple with a scattering of long russet plumes. It has a yellow-brown bill, reddish-brown legs and yellow iris. Juveniles have plainer brown plumage, less streaking on the upperparts and brown bill and legs. Although more solitary, it congregates in large colonies to nest.

Where to see: Occurs throughout the region in typical heron open habitat of rice fields, swamps, mangroves and lakes, and often inland.

Great Egret *Ardea alba* 80–104cm
Indonesian name: Kuntul Besar

The largest of the white egrets, with a more substantial bill and a peculiar kink midway along the neck. During the breeding period, the bare facial skin becomes greyish-blue and the lores pale green, the bill turns black and the bare thighs turn red. At other times of the year, the facial skin is dull yellow, the bill is yellowish, occasionally tipped black, and the feet and legs are all black. Juveniles have a black-tipped, yellow bill.

Where to see: Occurs throughout the region, preferring mudflats, mangroves and coastal marshland, as well as rice fields, where it is found either alone or in small groups, feeding on small fish and frogs.

Intermediate Egret *Ardea intermedia* 64–72cm

Indonesian name: Kuntul Perak

It is intermediate in size between the Great Egret and Little Egret but is less common. Recognised by its overall slightly off-white plumage, no head plumes but long back and chest plumes, and bare yellow facial skin only to the front of the eye. Unlike the Little Egret it has a yellow bill tipped black.

Distinguished from the larger Great Egret by its smooth, unkinked neck.

Where to see: Occurs throughout the region. It can be found in rice fields, mangroves and swamps, and on coastal mudflats, but probably more often inland.

non-br.

Little Egret *Egretta garzetta* 55–65cm
Indonesian name: Kuntul Kecil

Larger and slimmer than the Cattle Egret, it has greyish-black legs and a black bill. Normally brilliant all-white plumage, in the breeding season it develops long pendulous plumes on the breast, back and nape and a small, dirty-yellow bare facial patch. The Asiatic migrant subspecies is distinguished by having yellow toes.

Where to see: Occurs throughout the region, mainly in mixed egret flocks in coastal lowlands, rice fields, on riverine mudflats and sand bars, and alongside small streams. It mixes freely with other herons and egrets in their breeding colonies and at night-time roosts.

Pacific Reef Egret *Egretta sacra* 58–66cm
Indonesian name: Kuntul Karang

The Pacific Reef Egret possesses two colour forms. The commoner form has uniform grey plumage, a short, thick neck, very pale chin, dark yellow bill and greenish legs. Less common is the white form, which resembles the Cattle Egret but is considerably larger and has a sleeker head and neck. It also has relatively short greenish legs and a dirty yellow bill.

dark morph

Where to see: Occurs throughout the region, generally confined to coasts and seen resting on rocks and cliffs when not hunting at the water's edge. More commonly found on the sandy beaches and reefs of offshore islands.

Cattle Egret *Ardea ibis* 45–55cm
Indonesian name: Kuntul Kerbau

A smallish white heron, sometimes tinged orange on the forehead. In breeding plumage it develops an orange wash over the breast, head and neck, as well as yellow-orange bill, legs and lores. Its more rounded body, short neck and short, thick bill distinguish it from other egrets. It nests and roosts colonially.

Where to see: Common throughout the region, found in rice fields and freshwater swampy areas. It is also attracted to grazing cattle, feeding on the insects they disturb. In Petulu village, Bali, a large nocturnal roost is protected by the belief that the birds carry the souls of Balinese killed in the 1966 anti-Communist slaughter.

Javan Pond Heron *Ardeola speciosa* 45cm
Indonesian name: Blekok Sawah

This heron normally appears dark brown, streaked light brown, with buffish-brown tail and underparts, and white wings, but in breeding plumage the neck and head become golden-buff and the back blackish. It has a black-tipped yellow bill and dull green feet.

Where to see: Occurs throughout the region in both inland and coastal areas. Common on freshwater marshes and rice fields.

Black-crowned Night Heron *Nycticorax nycticorax* 60cm

Indonesian name: Kowak-malam Abu

This stocky-bodied heron has a distinctive adult plumage of black back and crown, white neck and underparts, and grey wings and tail, with two long white plumes emerging from the nape. Both the bill and the neck are comparatively short. In breeding plumage, the normally dirty yellow legs become red. Females are slightly smaller and juveniles are streaked and mottled brown.

Where to see: Occurs throughout the region, often near freshwater wetlands, rice fields, swamps and mangroves. Also on mudflats.

juv.

Striated Heron *Butorides striata* 35–48cm
Indonesian name: Kokokan Laut

imm.

Adults are mainly dark grey, with a very dark greenish crown and long crest feathers. They have a pale buff neck and facial markings. Immatures appear stockier and are brown with a streaked breast and mottled upperparts.

Where to see: Common throughout the region's coastal wetlands, it prefers to remain close to cover, but is occasionally seen on rocky shorelines and exposed reefs.

Malayan Night Heron *Gorsachius melanolophus* 47–50cm
Indonesian name: Kowak Melayu

This solidly built heron has chestnut upperparts finely barred and speckled across the back and wings, and a pale chestnut neck and head with black trailing crest. Its underparts are white and flecked heavily with chestnut, buff and black. The throat is white, streaked black along the median line. It often (but not always) has bright blue lores. It has olive-green legs, and the short, stubby bill is olive-brown with yellow on the lower mandible. Nocturnal and very timid.

Where to see: Occurs as a migrant throughout the region, preferring dense vegetation, particularly bamboo and reeds surrounding inland marshes

Glossy Ibis *Plegadis falcinellus* 55–66cm
Indonesian name: Ibis Rokoroko

Ibises

The only ibis in the region with all-dark plumage, it has blackish-chestnut coloration with green, purple and bronze iridescences. Immatures are dark brown, with buff streaking on the head and neck. Legs are dark brown to olive-grey, and the long, decurved bill is brownish-grey.

Where to see: Occurs on Java and Bali, usually in small groups in rice fields, marshland and lake fringes. It often associates with egrets and herons, both at night-time roosts and at breeding colonies.

Black-headed Ibis *Threskiornis melanocephalus* 65–75cm

Indonesian name: Ibis Cucukbesi

The almost all-white body is offset by a black head and neck and a strong, black bill. The long legs are also black. The decorative plumed feathers on the tail are pale grey, and the few fine decorative neck feathers are white.

Where to see: Occurs rarely in Sumatra and north Java, in freshwater marshes and by lakes and rivers, flooded grasslands and rice fields, taking amphibians, crustaceans and small fish. Occasionally associated with grazing cattle.

Black-winged Kite *Elanus caeruleus* 30–37cm

Indonesian name: Elang Tikus

Named for its black shoulder patches and the fact that it often hovers like a kite before dropping on its prey. When in flight, this elegant little raptor might be mistaken for a gull. Its grey upperparts and white underparts, along with long black primaries and squarish tail, can be confusing. It has red eyes encircled by black.

Where to see: Occurs throughout the region, preferring open countryside and in dry areas of sparse woodland, at woodland edges, as well as in cultivated areas. Occasionally seen perching on exposed tree branches and telephone poles.

Black Baza *Aviceda leuphotes* 28–35cm
Indonesian name: Baza Hitam

Identified as a small, thickset, black hawk with a pointed long crest with white and reddish-brown wing patches. The underparts have blackish-brown bars across the upper breast becoming reddish towards the vent. The neck is black, the iris red-brown and legs bluish grey-black. In flight the primaries appear white, tipped black and the secondaries grey-brown. Juveniles are very similar but duller and browner.

Where to see: Occurs as a winter visitor to Sumatra and Java.

Javan Hawk Eagle *Nisaetus bartelsi* 55–60cm
Indonesian name: Elang Jawa

Indonesia's national bird, easily identified by its conspicuous white-tipped dark crest. It has a black moustachial stripe and crown, buff throat and underparts streaked black on neck and brown on belly, and white-feathered legs. The sides of the head and nape are a rich russet-brown, blending into dark brown upperparts. The longish tail is brown with black barring. The iris and legs are yellow and the cere dark grey.

Where to see: Endemic to Java. Found only in primary hill and mountain forests and open wooded areas in West Java, and on the east coast at Meru Betiri National Park. A last stronghold is Gunung Halimun National Park.

Crested Serpent Eagle *Spilornis cheela* 50–70cm

Indonesian name: Elang-ular Bido

Dark-coloured, the upperparts being a greyish chocolate-brown and the underparts similar, with shoulders, belly and flanks spotted white. It has a raised patch of white-spotted feathers on the top and back of the head, giving the appearance of a crest. In flight, the broad white tail-band and the band of white on the underwing-coverts are significant. It has yellow feet and a greyish-brown bill. Often detected by its loud, shrill call, *cwee-chee, chee-chee, chee-cheee*.

Where to see: Occurs throughout the region. Often seen over woodland and forests.

imm.

Besra *Tachyspiza virgata* 25–35cm

Indonesian name: Elang-alap Besra

Males have uniform dark grey-brown upperparts with a dark greyish-brown head, and a white throat with a black mesial stripe. The breast and belly are rufous-grey, heavily streaked with black and white in the centre of the upper breast. The lower belly is barred white, leading into white undertail-coverts. The pale-tipped tail shows three wide black bars. Females are generally browner. Sumatran birds (*A. v. vanbemmeli*) are more rufous. The bill is black, the cere grey and the iris yellow to orange-red.

Where to see: Occurs throughout the region but confined mainly to mountain and foothill woodland.

Owls

Eastern Barn Owl *Tyto javanica* 33–50cm

Indonesian name: Serak Jawa

The Barn Owl is widespread throughout the world and familiar to many. It has an obvious heart-shaped white facial disc and underparts that vary from white to brown with darker speckling. The head and upperparts colouration is also variable from light brown to grey. Recognised in flight by its generally pale colouration, long wings, short, squarish tail and feet that do not extend beyond the tail.

Where to see: Occurs throughout the region, albeit sparsely. It is nocturnal in Indonesia and absent from most of the mountainous or heavily forested areas.

Oriental Bay Owl *Phodilus badius* 23–29cm

Indonesian name: Serak Bukit

This small owl has rather large blackish eyes set back in a deep fawn-coloured facial disc bordered by a thin dark brown line of feathers. The short, buff ear-tufts appear as an extension to the top of the facial disc. The upperparts are from chestnut to fawn with small buff spots and underparts buff-pink with small black spots. The short wings are barred black, the bill is buff and feet buff-brown. Females are a little larger and juveniles paler and duller.

Where to see: Occurs on Sumatra and Java, mainly in lowland forests and sometimes approaching mangroves and cultivated areas.

normal morph

rufous morph

Collared Scops Owl *Otus lempiji* 20–24cm

Indonesian name: Celepuk Reban

Identification is not easy, as the species occurs in several colour forms. It can be dull reddish-brown to grey-brown, with the upperparts mottled and blotched with buff and black; the underparts are paler and streaked black. It has a pale sandy-buff neck collar and longish buff ear-tufts. Bill and feet are pale yellow-pink. The regular soft *whoop* call of the male and the *weeoo* and *plop* calls from the female are often heard alternately as a continuous duet.

Where to see: Occurs throughout the region, almost everywhere below 1,500m. Often seen hunting for insects and small rodents from trees and prominent perches in built-up areas of towns.

rufous morph

brown morph

Buffy Fish Owl *Ketupa ketupa* 40–44cm

Indonesian name: Beluk Ketupa

This owl has an amusing appearance, with bright yellow eyes and its conspicuous horizontal ear-tufts. Its upperparts are brown, streaked black and buff, and the underparts are bright rufous-buff with narrow black streaks. It has whitish eyebrows, a grey bill and yellow feet.

Where to see: A mainly nocturnal species occurring throughout the region and remaining hidden among trees during the day. At night, it ventures outside the forest into parks, rice fields and often alongside roads. Always close to water, on which it is dependent for its diet of frogs, fish and other aquatic life.

Javan Owlet *Glaucidium castanopterum* 23–25cm

Indonesian name: Beluk-watu Jawa

This owlet has a warm brown face barred pale orange and white eyebrows with no apparent facial disc. The head and shoulders are brown subtly barred pale orange and upperparts are warm chestnut. The wings and tail are barred yellow and brown, and the brown upper breast subtly barred orange-brown. The underparts, including breast, belly and flanks, are pale buff boldly streaked chestnut. The edges of the scapulars are white forming a line bordering the wing. The iris is yellow and the bill and feet are green-yellow.

Where to see: Occurs in primary and secondary forests on Java and Bali.

Brown Boobook *Ninox scutulata* 27–33cm

Indonesian name: Pungguk Coklat

Adults and immatures are very similar, having grey-brown heads and dark brown upperparts. The underparts are white, heavily marked with heart-shaped brown spots, and the barred tail has a grey-brown subterminal band and white tip. There is a white facial patch above the greenish-black bill, and the eyes are orange-yellow.

Where to see: Occurs throughout the region, commonly in primary and secondary forest and often in pairs.

Javan Trogon *Apalharpactes reinwardtii* 34cm

Indonesian name: Luntur Jawa

Similar to but larger than the Sumatran Trogon (*Apalharpactes mackloti*), this bird has blue-green, sometimes turquoise upperparts that extend from the nape to the rump; below, the tail is rich steel-blue. The wing-coverts are finely barred yellow and black, and throat and belly are yellow separated by a broad, pale grey-green breast-band. The crown is bright green and a broad blue orbital ring surrounds a black iris. The bill is bright red and the feet orange.

Where to see: Occurs mostly in the submontane forest of West Java.

Orange-breasted Trogon *Harpactes oreskios* 25–27cm

Indonesian name: Luntur Harimau

This trogon has a striking orange breast, paling to yellow at the vent, but above the breast it becomes dull olive-grey turning olive-yellow over the head. The dark eye has a blue orbital ring. The upperparts are reddish-brown becoming paler at the rump. The undertail feathers are white and black and the wing-coverts are black finely lined white. Sumatran males have a more yellow upper breast, yellow-grey in the female.

Where to see: Occurs in lowland and submontane forest of Java and Sumatra.

Great Hornbill *Buceros bicornis* 95–105cm

Indonesian name: Rangkong Papan

The enormous yellow bill and casque, black face, yellow-stained white neck and upper breast, and black subterminal bar on a white tail separate it from other hornbills of the region. The back, lower belly and wing-coverts are black, and a white wing-bar is often stained with yellow. They have prominent 'eyelashes', similar to other hornbills. Males are larger than females and have a red iris and a black orbital ring of bare skin. The female's iris is white and the bare skin is pinkish-red.

Where to see: An uncommon species, confined to lowland and submontane primary evergreen forests of Sumatra.

Oriental Pied Hornbill *Anthracoceros albirostris* 55–60cm

Indonesian name: Kangkareng Perut-putih

Recognised by its mainly black plumage with a white lower breast, belly and flanks, this species also has white patches behind the eye and at the base of the lower mandible. The huge horn-coloured bill is surmounted by a banana-shaped casque marked black at the front end. Females have more black on the bill. The tail shows variable amounts of black and white, particularly on the outer feathers. Normally found in small parties. Utters loud, raucous and cackling laughs and chicken-like cackling, *puk-puk-puk-puk*.

Where to see: Occurs in clearings and edges of more remote lowland and secondary forest, down to the coast.

Wreathed Hornbill *Rhyticeros undulatus* 75–85cm

Indonesian name: Julang Emas

Apart from the whitish, often dirty yellow, tail and the male's chestnut-crowned, creamy head, the plumage is black. The male has a bare yellow gular pouch, that of the black-headed female being blue. Like all hornbills, it relies on hollow trees when nesting, the male reducing the size of the entrance hole with mud while the female is incubating, and the female doing likewise after hatching until the young are ready to fledge.

Where to see: Common in lowland and hill forests of Sumatra, it is confined to a few remaining undisturbed forests on Java and Bali.

Kingfishers

Blue-eared Kingfisher *Alcedo meninting* 16–17cm

Indonesian name: Raja-udang Meninting

This woodland counterpart of the Common Kingfisher is very similar in its behaviour, moving between waterside perches in rapid flight. Blue-eared has a more contrasting appearance, its upperparts being a darker but shinier blue and its underparts a bright orange-red. The ear-coverts are blue, legs orange and the iris is dark brown. It dives into water to catch fish, which it kills by bashing it on the perch before swallowing it whole, head-first.

Where to see: Occurs throughout the region, where it frequents forest rivers, streams, lakes and occasionally estuarine habitat.

imm.

Common Kingfisher *Alcedo atthis* 16–17cm
Indonesian name: Raja-udang Erasia

This common kingfisher occurs all the way across Eurasia and to New Guinea. Upperparts are a shiny bright blue-green, with paler blue uppertail-coverts and a small patch in the centre of the back. The underparts and ear-coverts are rufous. There is a white stripe on the side of the neck. Feet are red, the bill black, the iris dark brown. The Bali subspecies *A. a. floresiana* has darker blue upperparts and a few blue feathers in the rufous ear-coverts. In flight, it has a drawn-out, high-pitched call, *tseeep, tseeep.*

Where to see: Occurs throughout the region, frequenting lowland open country with freshwater mangroves, rivers and lakes.

Stork-billed Kingfisher *Pelargopsis capensis* 35cm
Indonesian name: Pekaka Emas

The largest kingfisher in the region, with an enormous red bill. Easily identifiable. The back, wings and tail are blue-green and the underparts pinkish-orange. The head is brown, becoming paler around the neck. It has short red legs. It feeds on fish, crabs and amphibians.

Where to see: Occurs throughout the region, inhabiting coastal mangroves and lowland rivers and marshland, as well as cultivated land and woodland adjacent to rivers. It is now an uncommon resident, and possibly extinct on Bali.

White-breasted Kingfisher *Halcyon smyrnensis* 27–29cm
Indonesian name: Cekakak Belukar

The white throat and 'bib' contrasting with the chocolate-brown head and underparts are distinguishing features of this large kingfisher. The upperparts, tail and upper surfaces of the wings are iridescent mid-blue, and there is a patch of brown on the upperwing-coverts. A loud screaming call, 'kee-kee-kee-kee', announces the bird's presence.

Where to see: Occurs in Sumatra and West Java, by rivers and ponds in mainly lowland open areas and even along the coast.

Javan Kingfisher *Halcyon cyanoventris* 27cm
Indonesian name: Cekakak Jawa

This dark-looking kingfisher is dark brown on the head, with a lighter brown collar and upper breast. The belly and hind collar are purplish-blue, and the back and wing-coverts dark purple. Flight feathers and tail are bright blue. The white wing patches are particularly conspicuous in flight. Both bill and feet are red. It is not reliant on fishing for its food, having adapted well to a diet of insects and terrestrial prey.

Where to see: Endemic to Java and Bali. Often seen on an isolated perch in open grassland, it remains near water where prey items are more abundant.

Sacred Kingfisher *Todiramphus sanctus* 22cm

Indonesian name: Cekakak Australia

The upperparts are green tinged grey, including the forehead, crown, mantle and back. The rump is turquoise and the tail blue-green, darker on the outer feathers. The throat, lower cheeks and collar are warm buff to white above which passes a black mask reaching narrowly across the nape. There is a buff to orange loral spot. The flight feathers are black and upperwing-coverts blue-green. The breast is buff becoming pale chestnut on the undertail-coverts. The legs are black and bill blackish above and buff below at the base.

Where to see: A migrant frequenting the mangroves, small lakes and coastal regions of Java and Bali, occasionally to east Sumatra

Collared Kingfisher *Todiramphus chloris* 23–25cm ✓
Indonesian name: Cekakak Sungai

At first sight this species is bright blue and white, but at close range the blue of the crown, back, wings and tail has a brilliant green iridescence. It has a strikingly clean white collar and underparts. The lores are white, and a black stripe passes through the eye and around the back of the head. A very common kingfisher, as is evident from its continual calling throughout the day. Its harsh, noisy *chue-chue-chue-chue-chue* carries over long distances.

Where to see: Occurs throughout the region. It prefers open country around the coast and wherever there is water in which to fish. *Komodo Island.*

Blue-tailed Bee-eater *Merops philippinus* 30cm
Indonesian name: Kirik-kirik Laut

The upperparts of this colourful bee-eater are green from the crown to back leading to a beautiful pale-blue rump and tail. The underparts are blue-green from below the yellow chin to the blue undertail-coverts. It has a wide black eye-stripe outlined by a thin blue line above and below. The throat and cheeks are red tinged brown. The bill is black, the iris ruby-red and feet grey-brown.

Where to see: Occurs throughout Sumatra and Java. More often found in open areas near water.

Common Dollarbird *Eurystomus orientalis* 27–30cm

Indonesian name: Tiong-lampu Biasa

Overall, it appears dark bluish-brown with an upright stance, with a large head, solid orange-red bill and blue throat. In flight, two circular light-blue contrasting patches on the underwing-coverts are evident, giving the bird its English name. Immatures are duller and lighter with blackish bills.

Where to see: The only member of the roller family to be seen throughout the region. Widespread but not common; usually singly or in pairs. Hunts from exposed perches in open country by swooping on its insect prey.

Coppersmith Barbet *Psilopogon haemacephalus* 15–17cm

Indonesian name: Takur Ungkut-ungkut

A drawn-out, resonant *tonk-tonk-tonk-tonk* call, reminiscent of a vibrating hammer on metal, announces this bird's presence. The back, wings and tail are green, with fawn underparts streaked black. Those found on Java and Bali have a red crown, chin, throat and cheek; those on Sumatra have the throat, cheek and eyebrow yellow. A thin black necklace separates the red and yellow neck markings. The legs are red.

Where to see: Occurs throughout the region. Frequently found in trees in open country and dry deciduous forest. It also ventures into town gardens and parks.

Blue-eared Barbet *Psilopogon australis* 16cm

Indonesian name: Takur Tenggeret

Two plumage variations occur in the region, both having green body plumage, a black malar stripe and breast stripe, and a blue chin and crown. Javan birds have patches of yellow on the cheek and below the black breast stripe, whereas the birds on Sumatra lack the yellow but have red cheeks. Also known as the Yellow-eared or Little Barbet, it is more likely to be heard than seen. The calls are a fast, repeated *tuk-trrk, tuk-trrk* and a series of high-pitched whistled or trilled notes.

Where to see: A common barbet of forests and plantations, occurring from the coast to lower hills throughout the region.

Fire-tufted Barbet *Psilopogon pyrolophus* 28–29cm

Indonesian name: Takur Api

As with most barbets, the main plumage is bright green. It has a yellow band across the breast, outlined by a black line below. The rear crown and nape are red, the chin green, the ear-coverts grey and the eye-stripe and forehead black. It has a prominent, red-tipped, black hairy tuft at the base of the lores. It is very fond of ripe fruit, particularly small figs, and consumes insects. It will perch stationary high in the trees for long periods while frequently uttering its harsh buzzing call, very reminiscent of a cicada.

Where to see: Occurs on Sumatra and West Java, where it prefers tall, older forest trees in the lower hills and on mountain slopes.

Black-banded Barbet *Psilopogon javensis* 26cm

Indonesian name: Takur Tulung-tumpuk

Quite large, with green body plumage and tail, this barbet has a yellow crown and a yellow spot below the eye. It also has a red throat, under which is a solid black collar which joins a black stripe running through the eye. The feet are pale olive-green and the bill black. Like with other barbets, the feet are adapted for climbing tree trunks, having two toes pointing forward and two back. It is mainly a fruit-eater with a penchant for small, ripe figs, seeds, buds, flowers and occasionally insects.

Where to see: Occurs around clearings and forest edges in the more open lowland and hill forests of Java and less so on Bali.

Flame-fronted Barbet *Psilopogon armillaris* 20–22cm

Indonesian name: Takur Tohtor

This robust little barbet has predominantly green upperparts, slightly paler underparts and a prominent yellow-and-orange forehead. The rear of the crown to the nape is blue and there are small patches of orange either side and just below the front of the neck. The strong bill is black, as is the frontal eye-stripe. Birds from East Java and Bali appear larger with darker orange on the forehead and lower neck.

Where to see: Occurs on Java and Bali at edges of lowland and hill forest; occasionally around nearby habitation.

Lineated Barbet *Psilopogon lineatus* 25–30cm

Indonesian name: Takur Bultok

This barbet has a green back, wings and tail, and a brown head and upper breast heavily streaked with buff. The remaining underparts are green, also streaked with buff. It has a broad orange-yellow orbital ring of bare skin. The heavy bill is variable in colour, from pale pink to yellow, and the feet are yellow.

Where to see: Occurs only on Java and Bali, frequenting sparse, dry coastal woodlands, edges of savanna, acacia forest, orchards and cultivated areas; often gathers with other barbets at fruiting trees.

Sunda Pygmy-woodpecker *Picoides moluccensis* 13cm

Indonesian name: Caladi Tilik

This little woodpecker has short legs and grey feet, with two toes pointing forward and two backward. Its upperparts are dark brown with white barring, giving a partially spotted appearance. The underparts are whitish with black streaking. It has a rich brown crown, black malar stripe, white face patch with a greyish-brown centre, and thin red stripe behind the eye.

Where to see: Occurs throughout the region. It prefers mangroves, open woodland and coastal secondary forest.

Common Flameback *Dinopium javanense* 28–30cm

Indonesian name: Pelatuk Besi

The male has a red crown extending into a crest at the rear, whilst the female's crown is a more flattened mass of black-and-white feathers. Otherwise, both sexes are similar, having a golden-yellow mantle and wing-coverts, red lower back and rump, black tail and primaries, and a white face with a black eye-stripe and single solid malar stripe. The white underpart feathers are edged black, forming a scaly pattern. It is unusual in having only three toes. Noticed by its continual 'churring' contact calls.

Where to see: Occurs throughout the region. A common lowland forest woodpecker, regularly appearing around cultivation, plantations and even in gardens, and usually in pairs.

Crimson-winged Woodpecker   *Picus puniceus*   25cm

Indonesian name: Pelatuk Sayap-merah

This predominantly red and olive-green woodpecker has a yellow hindcrown appearing as an extension to the long red crown and crest feathers. The malar stripe is red (only on the male) and lores black. The ear-coverts are olive-green running down the neck and onto the breast and back, with the upper breast and throat tinged brown. The wing-coverts are red, the bill brown above, yellow below. The iris is red and the legs are dirty green.

Where to see: Occurs commonly in primary and secondary forest on Java and Sumatra, often in mixed flocks.

Laced Woodpecker *Picus vittatus* 30–33cm

Indonesian name: Pelatuk Hijau

This woodpecker has green upperparts and nape, a yellow rump and a black tail. The underparts are buffish-yellow with the feathers edged in green, giving the laced appearance. Males have a red crown, this being black in females. The cheeks are blue-grey, bordered by a black eye-stripe and malar stripe, both of which are mottled with white. The black primaries are barred with white. The bill is black and the feet grey-green.

Where to see: Occurs throughout east Sumatra, Java and Bali. Prefers coastal woodland and forest, and often found in mangroves, bamboo thickets and coconut plantations.

Checker-throated Yellownape *Chrysophlegma mentale* 26–29cm

Indonesian name: Pelatuk Kumis-kelabu

This sombre dark green and red woodpecker has pale yellow nape feathers appearing as a nuchal crest. The throat is white and black chequered. A wide dull rufous band stretches from the rear of the ear-coverts on the sides of the neck and onto the breast. The wing-coverts are intense chestnut and flight feathers and uppertail brown-black. It has a long grey bill, a red-brown iris and grey legs.

Where to see: Occurs commonly in primary and secondary forest on Java and Sumatra, often in mixed flocks.

White-bellied Woodpecker *Dryocopus javensis* 40–48cm

Indonesian name: Pelatuk Ayam

A very conspicuous and striking appearance, being very large, with all-black plumage apart from a white belly. Males have a red crest and cheek patches. The long, pointed bill is black and the feet grey. Rather solitary, this woodpecker usually makes its presence known by its loud 'laughing' and 'barking' calls, as well as its heavy hammering on dead branches. It often calls in flight with a raucous, echoing *kiow, kiow, kiow*.

Where to see: Occurs throughout the region's open lowland forests

Black-thighed Falconet *Microhierax fringillarius* 14–17cm

Indonesian name: Alap-alap Capung

This tiny falcon has black upperparts but shows white spotting on the tail and secondaries. The crown is black, edged with a white patch above the bill and a white stripe behind the eye, and the sides of the face and ear-coverts are black. The chin and belly are rufous with white margins. The legs and cere are black and the iris brown. Females are a little larger than males, and juveniles have the white on the head tinged red.

Where to see: Occurs throughout the region. It frequents edges of forests and mangroves, and open country and scrub, and occasionally is seen over rice fields.

Indonesian Kestrel *Falco moluccensis* 26–32cm
Indonesian name: Alap-alap Sapi

The male has chestnut upperparts, heavily but clearly spotted dark brown and extending over the wings. The top of the head, nape and breast are more buffish-chestnut, with the spots more streaked. The blackish iris is surrounded by a yellow eye-ring and a patch of bare yellow skin. The cere is yellow, bill grey and legs yellow with black claws. The undertail-coverts are buff leading into a pale grey tail with a broad black subterminal band and white tip.

Where to see: Occurs on Java and Bali in open habitat, often close to habitation.

Oriental Hobby *Falco severus* 27–30cm
Indonesian name: Alap-alap Macan

The upperparts are blackish and underparts chestnut, darker on the chest and belly, becoming paler and buffer on the vent. The chin and throat are whitish but often light chestnut. The feet, cere and orbital ring are yellow. Juveniles have similar upperparts but have buff underparts heavily streaked brown. Colour variations arise across the range; darker on Bali to lighter on Sumatra.

Where to see: Occurs in open areas of woodland, grasslands and even rice fields of Bali, Java and east Sumatra.

Yellow-crested Cockatoo Cacatua sulphurea 33cm

Indonesian name: Kakatua-kecil Jambul-kuning

Similar to but smaller than the Sulphur-crested Cockatoo (*Cacatua galerita*). This all-white cockatoo has a lovely crest of stiff yellow feathers and yellow cheek patches. It has a black bill and dark grey feet. Its flight appears laboured, with alternating glides and fast-flapping wingbeats. Usually lives in pairs but often seen in small noisy groups, screeching, squawking and whistling, and raising its crest as it calls. Widely exploited by the pet trade.

Where to see: Occurs at forest edges on Nusa Penida, Bali. Scarce.

Red-breasted Parakeet *Psittacula alexandri* 33–37cm

Indonesian name: Betet Biasa

The upperparts from the nape to rump, including the wings, are deep green, with the median wing-coverts tinged brighter green. The head is grey-blue with a black line across the forehead to the eye. Another broader black line stretches across the chin and neck, and narrows below the cheeks. The underparts are bluish-pink immediately below the collar, shading to pink on the belly then to turquoise and to green on the undertail-coverts. The uppertail is bright blue. The bill is red, tipped horn-yellow, and feet are pale grey.

Where to see: Occurs on Java and Bali, where it prefers damp woodland in lowlands.

Blue-rumped Parrot *Psittinus cyanurus* 18–20cm

Indonesian name: Nuri Tanau

This small parrot has a blue crown and nape with the sides of the face blue, often tinged green-buff. The mantle and upper back are dark purple-grey to blackish. The lower back, rump and tail-coverts are bright blue and the tail is green to yellow on the outer feathers. The wing is green tinged maroon at the shoulder and underparts are grey-olive, becoming red on the flanks and underwing-coverts. The upper mandible is red and lower dark horn. Legs are grey-green and the iris is white. The female has a brown head.

Where to see: Occurs in lowland forest on Sumatra.

Sunset Lorikeet *Trichoglossus forsteni* 28–30cm
Indonesian name: Perkici Dada-merah

This is a brilliant multicoloured parrot with a very dark purplish-blue head. Apart from the pale green collar, the upperparts are dark green. It has red underwing-coverts and a yellow patch under the shoulder. The breast is red and belly dark blue, and the thighs show alternating green and yellow narrow bands. The feet are grey and the bill is red.

Where to see: A resident of Bali, normally seen in small, noisy, foraging parties flying low over the montane forest canopy.

Blue-crowned Hanging Parrot *Loriculus galgulus* 12–13cm
Indonesian name: Serindit Melayu

This small parrot is mainly bright green, darker on the wings and lighter below. Just below the nape is a small area of orange-yellow and similarly just above the bright red rump and uppertail-coverts. There is a small dark blue patch on the top of the crown and a bright red throat and upper-breast patch. The legs are orange and the bill is black. Females lack the red throat.

Where to see: Occurs on Sumatra and West Java, where it prefers lowland forest edges and plantations.

Javan Hanging Parrot *Loriculus pusillus* 12cm

Indonesian name: Serindit Jawa

This very small, mainly leaf-green parrot is well camouflaged, with brighter yellow-green underparts; the colour blends well with foliage. Males have a large yellow throat patch; females smaller. The feet and bill are orange. Most often seen in the treetops, often hanging upside-down to feed on flowers and small fruits. They roam around in small flocks, often making high-pitched shrieking calls.

Where to see: Endemic to Java and Bali. Inhabits lowland swamp forest, particularly areas with flowering *Casuarina* trees. Common.

Green Broadbill *Calyptomena viridis* 14–17cm

Indonesian name: Madi-hijau Kecil

This broadbill is all green apart from the male's obvious black spot on the cheeks below the ear-coverts and broad, black-barred markings on the wings. It has a peculiar feather tuft on the forehead, a short, black bill, short tail and broad, pointed wings with black flight feathers. A cream-coloured eye-ring surrounds a dark brown iris. The legs are buffish-green. Females and juveniles are paler and lack the heavy black markings.

Where to see: Occurs on Sumatra. Generally rainforest dwelling but also in swamp forest and riverine habitats.

Long-tailed Broadbill   *Psarisomus dalhousiae*   23–26cm

Indonesian name: Madi Injap

Distinctively marked and brilliantly coloured, this slender, long-tailed broadbill is particularly attractive. With green upperparts and green-turquoise underparts, it has a black skullcap with a blue crown patch and bright yellow spot behind the ear-coverts. The face and throat are bright yellow, leading to a narrow, almost complete pale yellow collar. The flight feathers are black and blue, as is the long tail. The iris is grey-green, legs yellow-green and bill blue-green.

Where to see: Occurs in primary and secondary evergreen and broadleaf forests of Sumatra.

Silver-breasted Broadbill *Serilophus lunatus* 16–17cm

Indonesian name: Madi Dada-perak

The male has a pale grey-brown head with a light grey forehead and a broad black supercilium. The upperparts are grey at the collar blending to chestnut at the rump and uppertail-coverts. The breast and belly are white. The flight feathers are brilliant blue and black and the tail black. There is a narrow orbital ring around a dark blue-green iris. The bill is pale blue-green, sometimes showing orange at the base, and the legs are dirty yellow.

Where to see: Occurs sparsely in submontane forests of Sumatra.

♀ ♂

Banded Broadbill *Eurylaimus javanicus* 21–23cm
Indonesian name: Sempur-hujan Rimba

This prettily coloured broadbill has a black mantle, back and uppertail-coverts coarsely marked with yellow patches. The flight feathers are brown-black and tail and wing-coverts black. The underparts and head are maroon-red, becoming lighter on the belly. A thin blackish band stretches across the upper breast. The bright cream iris is surrounded by an area of black from the lores to above the ear-coverts. The bill is pale blue tinged green and the legs are grey-pink.

Where to see: Occurs in Java and Sumatra's evergreen and mixed deciduous forest near wetlands.

Black-and-yellow Broadbill *Eurylaimus ochromalus* 13.5–15cm
Indonesian name: Sempur-hujan Darat

This brightly coloured broadbill has a black head covering the nape, ear-coverts and throat below which is a broad white collar. Below that is a black breast-band, unbroken in the male, broken centrally in the female. The underparts are pink, becoming yellow at the undertail-coverts. The tail is black, the bill is bright blue tinged green and black at the edges, the iris yellow-white, and legs fawn to blue. It has a black mantle, back and uppertail-coverts coarsely marked with yellow patches.

Where to see: Occurs on Sumatra. Prefers lowland forests.

♀ ♂

Javan Banded Pitta *Hydrornis guajanus* 20–23cm

Indonesian name: Paok Jawa

A beautiful ground-living bird with a distinctive plumage, including a black head with a long, broad yellow eyebrow. The upperparts and wing-coverts are plain brown, the rump and tail blue. The wings are dark brown and have a thin white wing-bar. The chin and throat are yellow-white bounded below by a dark blue band.

The remainder of the breast and underparts are finely barred yellow and black. Legs are pale purple-grey, the bill is black and iris very dark brown.

Where to see: Occurs on Java and Bali, inhabiting mainly lowland forest and shaded dense thickets

Asian Hooded Pitta *Pitta sordida* 16–18cm

Indonesian name: Paok Hijau

This bird hops around foraging for insects among the leaf litter and rotting wood on the forest floor. It is a plump, medium-sized pitta with long legs. Residents have green plumage with a black head, blue-grey wing-coverts, and crimson undertail-coverts and lower belly. The bill is black, and the legs are pinkish. Migrants (*P.s.cucullata*) have a dark-brown cap and more black on the belly. If disturbed, it flies close to the ground with rapid wingbeats, revealing conspicuous rounded white patches on the wings.

Where to see: Occurs in the swamp forests of Sumatra and Java.

P. s. cucullata

P. s. mulleri

Blue-winged Pitta *Pitta moluccensis* 18–20cm
Indonesian name: Paok Hujan

This very colourful pitta is distinguished by its bright blue wing patch. The primary wing feathers are black with broad white tips. The upperparts are green over the mantle and turquoise-blue at the rump. A short, black tail is tipped blue-green. The throat is white and underparts pale cinnamon. A bright red central belly extends to the undertail-coverts. The pale chestnut crown and supercilium contrast with a broad black stripe extending from the lores across the nape. The iris is brown and bill black. Feet are pinkish.

Where to see: A non-breeding winter migrant to Sumatra, occupying a variety of habitats, from forests to city parkland.

Wallace's Elegant Pitta *Pitta concinna* 19cm
Indonesian name: Paok Wallace

Wallace's Elegant Pitta has chin and head black to the nape, with an orange head stripe extending lengthways above the eye from the lores and becoming white at the nape. The upperparts are green tinged brown. The rump is blue and the black tail is green-tipped. The flight feathers are black with the green wing-coverts showing a blue shoulder patch. The underparts are orange-buff with a central black streak leading to a red vent. The bill is black, the iris brown and the feet pale red-brown.

Where to see: Occurs in woodland and scrub on Nusa Penida, Bali.

Brown Honeyeater *Lichmera indistincta* 15cm

Indonesian name: Isap-madu Australia

Upperparts dark grey tinged olive, showing grey at the crown. Underparts are whitish on the throat becoming light grey on the breast and buff lightly tinged olive towards the undertail-coverts. Flight and tail feathers are grey edged yellow-olive. There is a blackish loral stripe and a ring of dark grey bare skin, broader behind the eye, which surrounds a dark-brown iris. The bill is black and legs are dark grey.

Where to see: Occurs in mangrove scrub and forest on Bali, often close to human habitation.

Golden-bellied Gerygone *Gerygone sulphurea* 9–10cm

Indonesian name: Remetuk Laut

This is the only Australian fairy-warbler in the region. Its upperparts are brown-tinged grey and its underparts bright yellow. The chin and neck are white, and it often shows a whitish loral spot. The dark grey-brown tail is tipped black, with a subterminal line of white spots. Its bill and legs are dark grey. Juveniles are paler generally and have a pale eye-ring.

Where to see: Resident throughout the region, occurring in open forest, bamboo and conifer thickets, mangroves and even plantations.

Small Minivet *Pericrocotus cinnamomeus* 14–16cm

Indonesian name: Sepah Kecil

The head, mantle and upper back are grey, with rump, vent and outer tail orange. The long central tail feathers are black, and the black wings have a prominent yellow-orange patch. Males have a black throat and red to orange belly; females are paler, with a whitish throat and breast.

Where to see: Occurs on Java and Bali, where it inhabits more open lowland forest and also trees and copses around cultivated land, gardens and mangroves.

Sunda Minivet *Pericrocotus miniatus* 20cm

Indonesian name: Sepah Gunung

The male's head, neck, throat, mantle, uppertail-coverts and lesser upperwing-coverts are black with a blue sheen. The black wing primaries are partially banded scarlet-red and along with the red greater upperwing-coverts give the impression of an extended wing-bar. The breast, belly, lower back and tail-coverts are also scarlet. Females have a red supercilium, red chin and cheeks, and reddish mantle.

Where to see: Occurs in the montane forests of Java and west Sumatra.

Scarlet Minivet *Pericrocotus flammeus* 19cm

Indonesian name: Sepah Hutan

The male's head and mantle are jet-black with a blue sheen, and the underparts, rump and outer edges of the tail are red. The mainly black wing has two distinct red patches. The female has a grey crown and upperparts, with yellow to orange outer tail and underparts including chin, forehead and ear-coverts, but leaving a suggestion of a grey eye-stripe. She has two yellowish wing patches, and her rump is grey-green.

Where to see: Occurs throughout the region. A forest-dwelling bird, usually seen foraging in the treetops in small groups.

Javan Cuckooshrike *Coracina javensis* 28cm

Indonesian name: Kepudang-sungu Jawa

The male has generally grey plumage apart from the white tip to the blackish tail and white edges to the dark grey wings. The grey on the upper belly pales to almost white at the vent. The ear-coverts and face are dark grey to black. Females are similar but paler, with grey-barred underparts and flanks. It should not be confused with the very similar but smaller Sunda Cuckooshrike (*C. larvata*), which occurs alongside this species at some elevations.

Where to see: Occurs in more open forested and grassy regions of Java and Bali.

♂

Lesueur's Triller *Lalage sueurii* 17–18cm

Indonesian name: Kapasan Sayap-putih

Males are predominantly black and white and females brown above and buff below. The top of the male's head and upperparts have greenish iridescence. There is a thin white supercilium under which a black eye-stripe passes through the eye to meet the black nape. The rump and lower-tail coverts are grey and the white upperwing-coverts form a white panel. The underparts, including the throat and ear-coverts, are white.

Where to see: Occurs in Central and East Java and Bali, usually in open forest and often occurring close to human settlements.

♂

Pied Triller *Lalage nigra* 16–18cm

Indonesian name: Kapasan Kemiri

The truly pied male is black from forehead to crown and nape and over the mantle, then grey on the uppertail-coverts. A black eye-stripe passes from the lores to the nape, over which is a broad white supercilium. The head is white below the eye-stripe along with the chin and remaining underparts. The wing feathers are black apart from a broad white patch across the median and greater coverts. The secondaries are black bordered white. Females have a brown crown and mantle and buff, lightly barred brown breast.

Where to see: Occurs throughout the region's coastal mangrove and open scrubland.

Lesser Cicabird *Lalage fimbriata* 19–21cm

Indonesian name: Kepudang-sungu Kecil

This small cuckooshrike has dark grey plumage, darker above and paler on the rump. The lores, ear-coverts and wings are blackish and the undertail-coverts pale grey. The bill and legs are black and the iris brown. Females are paler, with black barring on the underparts.

Where to see: Occurs throughout the region. Found in lowland forest, swamp forest and cultivated land, where feeds on insects.

Trilling Shrike-vireo *Pteruthius aenobarbus* 12cm

Indonesian name: Ciu Kunyit

Males of this small shrike-babbler have olive-green tinged yellow upperparts, including crown, ear-coverts and the outer parts of the secondary wing feathers. The upperwing-coverts are black tipped white, forming double wing-bars. The chin, throat and lower forehead are reddish-black with a narrow yellow band below the forecrown. It has a white eye-ring underlined black at the rear. The belly is bright olive-green, as are the ear-coverts, which are grey bordered and join with a grey supercilium. The legs are pink and the bill blue-grey. Females are duller with a buff-green throat.

Where to see: Occurs in the lower montane forests of Java.

♀

♂

♂

Pied Shrike-vireo *Pteruthius flaviscapis* 14cm

Indonesian name: Ciu Jawa

This plump little bird appears to have a proportionately large head and thickish bill. Males have a black head with a conspicuous white eyebrow, grey mantle and back, and black wings and tail, with the primaries tipped white and the tertials tinged chestnut and yellow.

The underparts are whitish. Females have a grey head, less distinct eyebrow, and greyish-green upperparts.

Where to see: Occurs on Java, more often in the canopy of more open lower montane forests.

Tenggara Whistler *Pachycephala calliope* 17–19cm

Indonesian name: Kancilan Tenggara

The male has brilliant black, white and bright yellow plumage. The yellow nape patch is joined by a narrow yellow band to the breast, belly and vent, whilst the head is black except for a white chin and throat, leaving a narrow black band between the throat and belly. The upperparts are olive-green and the tail is almost black. Females have drab olive upperparts and dirty buff underparts, with the lower belly and undertail-coverts tinged yellow.

Where to see: Occurs in East Java and Bali in hill and montane forest, impenetrable woodland and second growth.

Dark-throated Oriole *Oriolus xanthonotus* 17–19cm

Indonesian name: Kepudang Hutan

The male has head, mantle and neck to upper breast all black. The remainder of the upperparts are bright yellow. The flight feathers are black fringed with pale yellow. The tail is black with undertail-coverts and tips of central feathers also yellow. The belly is white, heavily streaked black. The bill is a muted orange, legs grey and iris dark red. The female is duller with upperparts generally olive green and underparts white streaked black with yellow undertail-coverts.

Where to see: Occurs in lowland forest edges of Sumatra and Java.

Black-naped Oriole *Oriolus chinensis* 23–28cm

Indonesian name: Kepudang Kuduk-hitam

The male is a brilliant yellow, with a distinct black stripe passing through the eye and across the nape. The flight feathers and tail are black, edged yellow. The female is much duller. Juveniles show olive where adults are black, and they have black-streaked, whitish underparts. It adheres to the tree canopy while foraging for insects and soft fruit.

Where to see: Occurs commonly throughout the region's lowlands and hills, favouring secondary and open forest, parks and gardens, mangroves and coastal scrub.

Black-and-crimson Oriole *Oriolus consanguineus* 21–24cm

Indonesian name: Kepudang Melayu

Easily identified as a medium-sized oriole, the males having all-black plumage except from the crimson-red upper belly and a small crimson patch on the primary coverts. It has a distinctive pale blue-grey bill and blue-black legs. Females are all black, with the breast tinged chestnut and streaked with grey on the belly. It has a cat-like, mewing call.

Where to see: Occurs on Sumatra, mainly in the treetops of broadleaf evergreen and hill forests.

♂

White-breasted Woodswallow *Artamus leucorynchus* 18cm

Indonesian name: Kekep Babi

The only resident woodswallow in the region, showing similarities to true swallows in stance and gliding flight; easily distinguished by its stockier appearance, broad, triangular-shaped wings and squarish, unforked tail. Its heavy bill is grey, and the entire upperparts except for the white rump are dark slate-grey. The underparts are white. Feeds on insects caught in flight.

Where to see: A common bird of open spaces, particularly in the lowlands and hills, throughout the region.

Black-winged Flycatcher-shrike *Hemipus hirundinaceus* 14–15cm

Indonesian name: Jingjing Batu

Male's upperparts are all black with greenish iridescence from the nape to the tail, apart from a white rump with slight barring on the upper rump. The underparts, throat and sides to the neck are white. Bill and legs are black and iris brown. Females and juveniles are similar but more grey-brown, with immatures looking more mottled.

Where to see: Found throughout the region's lowland forest edges.

Rufous-winged Philentoma *Philentoma pyrhoptera* 16–17cm

Indonesian name: Philentoma Sayap-merah

One of the larger flycatchers, occurring in two colour forms. The male of the more common rufous-winged form has dull-blue upperparts down to the centre of the back and blue underparts to the centre of the belly. The lower belly and vent are buff-white. The secondaries, greater coverts and tail are chestnut and the primaries dark grey-brown. The legs are blue-grey and the bill is black and the iris red. The blue form is all dark blue-grey with paler lower belly and vent. The female's upperparts are brown shading into dark grey towards the top of the head. Underparts buffish on the breast to white at the vent.

Where to see: Occurs on Sumatra, mainly in lowland forest.

Ioras

Common Iora *Aegithina tiphia* 12–15cm

Indonesian name: Cipoh Kacat

The Sumatran subspecies has olive-green to bright green upperparts, a white rump, and yellow underparts. The wings are black, with distinct white and yellow edges and two clear wing-bars. On Java and Bali the subspecies has more subdued colouration, the upperparts being a paler olive-green. There is a clear yellow eye-ring. Both bill and feet are bluish-grey.

Where to see: Occurs throughout the region, inhabiting open lowland forest and mangroves, but can frequently be found in town gardens.

♀

♂

Green Iora *Aegithina viridissima* 12–15cm

Indonesian name: Cipoh Jantung

The upperparts of the male are dark olive-green with the underparts paler, becoming yellower at the vent. The tail and flight feathers are blackish, sometimes edged olive-yellow. The coverts are black tipped white, forming two conspicuous wing-bars. A narrow band of black skin surrounds the eye, encircled by an area of pale yellowish feathers. The iris is brown and the legs are blue-grey. The bill is black above, blue-grey below. Females and juveniles are yellower and brighter.

Where to see: Occurs commonly in the canopy of Sumatran forests.

Sunda Pied Fantail *Rhipidura javanica* 18–19cm

Indonesian name: Kipasan Belang

This very active, small fantail displays a typical wing-drooping posture and regular flicking and fanning of the tail. The upperparts are dusky-black, the underparts white with a diagnostic black chest band, and there is a thin white eyebrow. When the tail is fanned, the broad white tips to the feathers are obvious. Both bill and legs are black and the iris is dark brown. Sumatran birds appear to have blacker upperparts and slightly longer tails.

Where to see: An open-habitat species occurring throughout the region's scrublands, open woodland, second growth, mangroves and even gardens.

White-bellied Fantail *Rhipidura euryura* 18cm
Indonesian name: Kipasan Bukit

Males and females have very
similar plumage; predominantly
dull-grey overall but below the
mid-belly it is white. There is
a broad white supercilium and
the outer tail feathers are tipped
white. Bill and legs are black
and the iris dark brown.

Where to see: Occurs in Java's
dense montane forests, often in
small groups with other species,
foraging for small flying insects.

Black Drongo *Dicrurus macrocercus* 25–30cm
Indonesian name: Srigunting Hitam

This drongo has black plumage with
slight iridescence and a long, deeply
forked tail with outer feathers which
tend to turn outwards at the tips.
Juveniles show pale grey barring on the
underparts. Unlike most other drongos,
favours open country and often
associates with cattle, which disturb
insect prey.

Where to see: Occurs on Java and
Bali. Frequently seen in open scrub,
cultivated land and rice fields, perching
on fencing, isolated branches and
telephone wires alongside roads.

Ashy Drongo *Dicrurus leucophaeus* 26–29cm

Indonesian name: Srigunting Kelabu

This drongo has the typical long, forked tail and upright stance. Its plumage is light blue-grey, with a tuft of black feathers at the base of the upper mandible. The subspecies in the region all vary in their grey colouration. Sumatran birds occasionally have a white loral spot, with those in north Sumatra being a little darker. Usually observed in pairs waiting on exposed branches to swoop on large passing insects.

Where to see: Occurs throughout the region. A bird of open woodland and forest edge and clearings.

Crow-billed Drongo *Dicrurus annectens* 27–30cm

Indonesian name: Srigunting Gagak

This sturdy but elegant drongo has
a strong, broad bill, well adapted to
taking aerial insects in flight. The base
of the bill is hidden by a layer of short
feathers. The head is black and the
remainder of the upperparts glossed
black with a bluish iridescence. The
underparts are black, often spotted
white. The long, blackish, forked
tail has the outer feathers curved
outwards. Legs and bill are black and
the iris dark red.

Where to see: Occurs as a migrant on
Sumatra and West Java, frequenting
wooded lowland and hills.

Javan Spangled Drongo *Dicrurus jentincki* 27–29cm

Indonesian name: Srigunting Jawaut

This unusual drongo has a peculiarly
fork-shaped tail, the outer feathers
spreading outwards and upwards
almost into a lyre shape. It also has an
extraordinary crest of long, hair-like
feathers from the front of the crown,
but this feature is variable. It has a very
pale buff eye. The general plumage
is glossy black with an iridescent
spangling, particularly on the crown
and upper breast.

Where to see: Occurs in East Java and
Bali. It is common in open areas of
lowland and submontane forests.

Lesser Racket-tailed Drongo *Dicrurus remifer* 25–28cm
Indonesian name: Srigunting Bukit

A spectacular, big, glossy-black drongo with a square-ended tail and extended outer tail-streamers almost 50cm long. Each streamer feather shaft terminates in an elongated oval web around 10cm long. A substantial tuft of short black feathers covers the base of the upper mandible, creating a long-headed appearance. It is smaller than the Greater Racket-tailed Drongo, has no frontal crest, and has no fork in the tail.

Where to see: Occurs in West Java and the Barisan Range and Batak Highlands of Sumatra. Inhabits dense montane forest and secondary forest, especially at edges.

Greater Racket-tailed Drongo *Dicrurus paradiseus*

30cm plus 35cm tail rackets   Indonesian name: Srigunting Batu

This large, forked-tailed drongo has beautifully extended outer tail feathers terminating in spirally twisted rackets on the outer edge of the shafts. Its forked tail easily separates it from the Lesser Racket-tailed Drongo. Plumage is glossy black, and adults often display a short frontal crest on the crown. The bill and legs are black.

Where to see: Occurs throughout the region. Usually found in pairs, often hawking for insects in glades in swamp, primary and secondary forests and in mangroves. It remains a common bird of the lowland forests of Sumatra.

Black-naped Monarch *Hypothymis azurea* 15–17cm

Indonesian name: Kehicap Ranting

A beautiful flycatcher with an azure-blue head and back. The male has a short, erect black hindcrown crest and a thin black band above and below the bill. A narrow black band extends

across the upper breast, with a grey belly paling to a white vent. Wings are grey, tail blue-black, bill bluish tipped black, iris dark brown and legs blue-grey. Females and juveniles are duller and greyer and lack the male's black markings. Small colour variations occur: Sumatran birds are more grey-blue, on Java and Bali males are pale blue and females have greyish-blue upperparts, and on Nusa Penida Island, Bali, females have a more bright-blue head.

Where to see: Occurs commonly throughout the region's forests, forest edges and scrubland.

Amur Paradise-flycatcher *Terpsiphone incei*

males: 20–25cm, females: 18–21cm Indonesian name: Seriwang Utara

Occurs in two colour forms: red and, much less common, white. In red form, upperparts and upperwing down to the tail are rich chestnut. The upper breast is pale grey shading to white at the belly and vent. The bill and legs are grey and the iris brown, encircled by pale blue bare skin. The red-form male has a long chestnut tail and complete glossy-black head, including a nuchal crest. The white form is similar to Blyth's Paradise-flycatcher but shows far less black on the edges of the primaries. The female's plumage is very similar, but has a much shorter tail.

Where to see: A scarce overwintering species of scrubland and forest edge, occurring on Sumatra and West Java.

Blyth's Paradise-flycatcher *Terpsiphone affinis* 20–30cm

Indonesian name: Seriwang Asia

Easily confused with Amur Paradise-flycatcher. It also occurs in two colour forms: red and, much less common, white. The red-form male has a complete glossy black head including a nuchal crest, whereas the upper breast to mid-breast the the flanks is dark grey before becoming white on the lower belly. The vent and tail are light chestnut, as are the remaining upperparts. The rarer white form is all white apart from the black head and some wing markings. The female's plumage is very similar but with a much shorter tail.

Where to see: A common over-wintering species of scrubland and forest edge, occurring on Sumatra and Java.

red-morph ♂

white-morph ♂

Jay Shrike *Platylophus galericulatus* 31–33cm

Indonesian name: Tangkar Ongklet

This bird is relatively inconspicuous until it starts to call, the harsh and rattling sounds revealing its presence. Its plumage is dark grey to black, apart from a broad white neck patch. Easily recognised by its tall, flat, upright crest. The bill is black and legs are dark grey.

Where to see: Occurs in Java and Sumatra's lowland broadleaf and hill forests, usually in small, noisy groups foraging through trees and bushes.

Tiger Shrike *Lanius tigrinus* 17–19cm

Indonesian name: Bentet Loreng

This little shrike has a robust, bluish-black, hook-tipped bill. The adult's crown is light grey extending well over the nape into the mantle, and it has a broad, mask-like eye-stripe from the lores over the ear-coverts (not present on immatures). The upperparts are russet-brown barred black down to the warm brown tail. Underparts, including chin to undertail-coverts, are white. The legs are grey-blue and iris dark brown. Females are duller with less obvious markings.

Where to see: Occurs in more open habitat and secondary forests throughout the region. Uncommon.

imm.

Long-tailed Shrike *Lanius schach* 20–25cm

Indonesian name: Bentet Kelabu

Easily identified by the long black tail, upright stance and brown, black and white plumage. It has a grey crown and nape, with chestnut mantle, back and rump; the underparts are clean white. Adults have a black mask from the forehead through the eye and across the ear-coverts. Juveniles tend to be duller, with dark grey head and nape and barring on flanks and back.

Where to see: Occurs throughout the region. Often in pairs, it roams through grassland and scrub, venturing into cultivated land and plantations up to the edge of human habitation.

Common Green Magpie *Cissa chinensis* 31cm

Indonesian name: Ekek Layongan

This magpie is predominantly green with a conspicuous black mask stretching through the eye to the nape. The more yellowish-green crown feathers also extend to form a short crest at the nape. The wing feathers are dark chestnut apart from the light green scapulars and upperwing-coverts. The upperside of tail is pale green; a red orbital ring encircles a brown iris; the substantial bill and legs are red. The bright green plumage is sometimes faded by sunlight to pale blue, particularly when in captivity.

Where to see: Occurs in the hill and mountain forests of Sumatra.

Racket-tailed Treepie *Crypsirina temia* 31–33cm including 18cm tail

Indonesian name: Tangkar Centrong

This member of the crow family has uniformly blackish plumage with a greenish-bronze sheen. It has black feet, a strong black bill and blue eyes. It differs from all similar birds in the region in having a long tail with spatulate tips to the central feathers.

Where to see: Occurs as a resident on Java and Bali. Usually inhabits open lowland and hill forests and plantations, and often in secondary growth, more cultivated areas, and even scrubland and gardens.

Sunda Crow *Corvus enca* 43–48cm

Indonesian name: Gagak Hutan

This entirely black crow is more slender and shorter-tailed than its larger cousin, the Large-billed Crow, which is also found in the region. It also has fewer throat hackles and inconspicuous bare black skin around the eye. Most of its plumage is slightly iridescent purple, except for the dull black underparts. The longish, slightly downcurved bill is black, the legs also black and the iris brown. In flight it shows a squarish tail and broad, stubby wings.

Where to see: Occurs in forest and at forest edges throughout the region, sometimes in pairs.

Large-billed Crow *Corvus macrorhynchos* 47–50cm

Indonesian name: Gagak Kampung

This bulky black crow is distinguished from its close relative, the Sunda Crow, by its strong, solid bill, more protruding forecrown and more pronounced throat hackles. Its overall plumage is glossed purple. The iris is dark brown and feet are black. Even the underparts show iridescence, except in juveniles, which are soft matt in appearance. In flight it has long and broad wings with wide-spread primaries and long tail.

Where to see: An open-country species occurring throughout the region, often close to human habitation.

Rail-babbler *Eupetes macrocerus* 28–30cm

Indonesian name: Sipinjur Melayu

The unique rail-babbler occurs in the region, where it is almost totally terrestrial and is mostly revealed by its sorrowful whistling call. It has a long neck and tail. Its plumage is mainly red-brown on both rear upperparts and underparts, which become redder on the belly, neck and underwing, and chestnut on the chin and throat. The forecrown is buff, the crown chestnut with a broad white supercilium above a broad black eye-stripe from the lores to the nape. Below the rear of the eye-stripe there is a bright blue patch over each gular sac. Bill and legs are black and the iris is brown.

Where to see: Occurs on Sumatra, primarily in dense forest.

Grey-headed Canary-flycatcher *Culicicapa ceylonensis* 12–13cm

Indonesian name: Sikatan Kepala-abu

This little flycatcher has a typical upright posture with a raised superficial crest. The head, from bill to nape, along with the sides of the face, neck and upper breast, is grey. The upperparts are dullish green, from the mantle to the rump as well as the upperwing-coverts. The black wing feathers are edged dull green. The underparts from breast to undertail-coverts are dullish yellow, becoming greyish on the flanks. A white eye-ring encircles a dark brown iris; the bill is black above, pink below and legs fawn colour.

Where to see: A common forest and woodland species, occurring throughout the region.

Cinereous Tit *Parus cinereus* 13cm

Indonesian name: Gelatik-batu Kelabu

These are relatively small, black, white and grey birds. The head and throat are black, with a black stripe from the throat to vent; the ventral stripe is wide on males, but on females can be very narrow and even broken. There is a large white patch on the cheek and a small white nape spot.

Where to see: Occurs throughout the region from lowland to open montane deciduous, coniferous and secondary forests as well as mangroves and even gardens. Usually seen foraging for insects.

Australasian Bushlark *Mirafra javanica* 13–15cm

Indonesian name: Branjangan Jawa

This bird occasionally perches in trees but is more often seen walking on the ground. It has dark-mottled, russet-brown upperparts, with paler underparts and white outer tail feathers. It has a broad buff supercilium and brown bill tinged yellow below. Its short, pinkish-grey legs have long claws. It has a weak undulating flight, and like many other larks it has a fluttering hover-flight in which it often sings, with a high-pitched trilling, before gliding slowly back to the ground.

Where to see: Common in open areas of short grass, rice field stubble and dry cultivated fields on Java and Bali.

Common Tailorbird *Orthotomis sutorius* 10–14cm

Indonesian name: Cinenen Pisang

The upperparts on this wren-like warbler are yellow-green from the nape to the tail, with the flight feathers brown edged olive. The head is bright chestnut from the lores to the nape, coloured paler on the sides of the face. The underparts from throat to the undertail-coverts are buffish-white, becoming grey-white on the flanks. The longish bill is brown above, pink below, the iris is pale brown and legs are pink-yellow. Females have slightly shorter tails and juveniles are duller.

Where to see: Common on Java, in more open scrub and gardens, even close to human habitation.

Ashy Tailorbird *Orthotomus ruficeps* 12cm

Indonesian name: Cinenen Kelabu

A small, plain-coloured tailorbird with a grey-brown back and greyish underparts, becoming white on the belly. Males have a rufous face, crown and throat, whereas females have a much paler rufous face and crown and white throat.

Where to see: This energetic little bird frequently occurs in the region's wetland areas in more open country and lowland forests, mangroves, bamboo thickets, coastal scrubland and well-vegetated gardens.

Olive-backed Tailorbird *Orthotomus sepium* 12cm

Indonesian name: Cinenen Jawa

This tailorbird has olive-grey upperparts, paler below with a whitish belly. The head of the male is almost completely light chestnut, including crown to below the ear-coverts and lores to nape; the chin being a little paler. Females have less chestnut on the face. The greyish tail has a fine whitish tip below a dark subterminal bar. The legs are pink, the quite long bill is brown above, pinkish below and the iris is pale red. Females are duller with a white chin.

Where to see: Occurs in Java and Bali's scrublands, cultivated areas and mangroves.

Deignan's Prinia *Prinia polychroa* 15–16cm

Indonesian name: Perenjak Coklat

This large, dull-coloured prinia has a substantial long brown tail. In breeding plumage it is duller still, with well-abraded plumage. At other times it becomes less dull with a whitish neck and supercilium in front of the eye, pale buff underparts and brown upperparts with dark streaking. The bill is black when breeding, otherwise brown to yellow-brown. Legs are pink-brown.

Where to see: Occurs on Java, preferring dry lowland scrub.

Bar-winged Prinia *Prinia familiaris* 13cm

Indonesian name: Perenjak Jawa

A noisy, versatile and gregarious little bird, often uttering its loud, high-pitched tweeting call as it searches for insects from ground to treetops. It is recognised by its long tail of black and white-tipped feathers, drab, olive-brown upperparts, and two distinctive white wing-bars. The white of the throat extends down the middle of the upper belly, with the flanks pale grey and lower belly and vent pale yellow.

Where to see: A very common endemic in the lowlands of Java and Bali, but less so on Sumatra, particularly in the north. It favours second growth, especially parks and gardens, mangroves, plantations and scrub.

Yellow-bellied Prinia *Prinia flaviventris* 12–14cm
Indonesian name: Perenjak Rawa

Appearing like a long-tailed warbler, it has olive-brown upperparts and tail, a bright yellow belly and a white chin, throat and upper breast. The head is greyish with a reddish-orange eye-ring and a very thin, and often broken, white supercilium. The legs and feet are dull orange, and the bill is dark brown above and pale brown below. This prinia has a weak song.

Where to see: Found on Sumatra and West Java, it is common in grassland, reedbeds and thick scrub in the lowlands. It has shy habits, with a tendency to remain hidden among long grass and reeds in wetland habitat.

Plain Prinia *Prinia inornata* 11cm
Indonesian name: Perenjak Padi

This small prinia has plain, grey-brown upperparts from the crown over the ear-coverts and back to its long, tapered, buff-tipped tail. The underparts are whitish tinged rufous towards the belly and vent. It has a broad white supercilium. The legs are pale brown. The bill is black in the breeding season, becoming pale brown later.

Where to see: Occurs on Java and Bali in a variety of habitats where scrubby grassland is present.

Zitting Cisticola *Cisticola juncidis* 10cm

Indonesian name: Cici Padi

This rather inconspicuous little brown bird is heavily streaked with dark brown and buff on its upperparts. The white underparts and vent are warm buff on the flanks, and the brown-and-black tail is tipped white. It has pink legs, a grey-black bill and a pale-brown iris. Its broad white supercilium distinguishes it from the Golden-headed Cisticola.

Where to see: Occurs throughout the region. Common in lowland grassland and reedbeds; it is attracted to wet habitat and rice fields.

Golden-headed Cisticola *Cisticola exilis* 9cm

Indonesian name: Cici Merah

Breeding males have a warm golden crown and rump. Non-breeding birds and females are very similar to the Zitting Cisticola but have a more golden head colour, and the buff supercilium is the same colour as the nape and head sides. The throat is white and the underparts buff. The dark brown tail is tipped buff.

Where to see: Occurs throughout the region. A common bird of Alang-alang grassland, rice fields and reedbeds on Java and Bali. Only locally common on Sumatra.

Oriental Reed Warbler *Acrocephalus orientalis* 19cm

Indonesian name: Kerak-basi Besar

This is a large brown warbler with a conspicuous pale buff supercilium underlined by a thin dark-brown eye-stripe. Its underparts are whitish, shading into buff on the flanks, rump and upper belly, with the upper belly and the sides of the breast sparsely streaked brown. The largish bill is brown above and tinged pink below, and the legs are blue-grey.

Where to see: Occurs throughout the region as a winter migrant. Usually found in lowland reedbeds and coastal marshes, but frequently occurs in rice fields and scrub near water.

Striated Grassbird *Megalurus palustris* 22–28cm

Indonesian name: Cica-koreng Jawa

Large, long-tailed warbler, the male being larger than the female but very similar in their buff-brown plumage. It has a long white supercilium above a dark brown eye-stripe. Upperparts are light brown, heavily streaked darker on the mantle. Underparts are white-buff with superficial streaking, darker on the flanks. The bill and iris are brown and legs yellow-brown.

Where to see: Resident in Java and Bali's wetlands and marshes where tall grass is present.

Pygmy Cupwing *Pnoepyga pusilla* 8–9cm

Indonesian name: Berencet Kerdil

This tiny, tail-less little babbler has warm, olive-brown upperparts marked only by the paler tips to the scapulars and wing-coverts. The underparts are brown with strong buff-white scaling. The face and ear-coverts are warm grey-brown, darker and less rusty-coloured overall in Javan birds.

The bill is grey-black, legs are light brown and iris is dark brown. Sexes are similar.

Where to see: Found mainly on the ground of montane forests of Java and Sumatra.

Barn Swallow *Hirundo rustica* 17–19cm

Indonesian name: Layang-layang Asia

The Barn Swallow is dark blue above, with a white breast and very long outer tail feathers. The forehead and throat are red, with a blue bar across the upper breast. Juveniles are duller and lack the tail-streamers.

Where to see: Overwinters in the region. It is more likely to be seen wheeling and gliding high in the sky, hunting aerial insects but often skims the surface of lakes and streams. Sometimes gathers in huge flocks to roost in reedbeds, tall grasses and even on city buildings.

Ruby-throated Bulbul *Pycnonotus dispar* 17–20cm

Indonesian name: Cucak Emas

This large bulbul has olive-tinged yellow-green upperparts and paler underparts. Distinguished by its black head with a superficial tufty crest, bright red throat and rounded tail with white-tipped outer feathers. Sexes are similar. Juveniles have a dark olive-brown nape. It has a black bill, blackish legs tinged pink and a pale orange iris.

Where to see: Occurs throughout the region in well-foliaged tall trees in open secondary forest and at forest edges.

Black-headed Bulbul *Microtarsus melanocephalos* 16–18cm

Indonesian name: Cucak Kuricang

Recognised mainly by its yellow-green body plumage and glossy black head and throat. The upperparts are greenish-yellow tinged brown, and the underparts brighter. The wings are black and the tail yellow, with a broad black subterminal bar. The bill is black, legs brownish-black and iris very pale blue. Juveniles are duller. Rare colour forms exist in which the yellow areas are replaced by dull grey or greenish-olive, but they have a yellowish-white vent and undertail-coverts, and whitish tip to the tail.

Where to see: Occurs throughout the region at forest edges, in riverine and coastal scrub and often along roadsides.

grey morph

yellow morph

Scaly-breasted Bulbul *Pycnonotus squamatus* 14–16cm

Indonesian name: Cucak Bersisik

The upperparts, including the mantle to the rump, are olive-green, the wings being a little darker and the black tail tipped white. The underparts are black on the breast and flanks with the feathers edged white, producing a scaled appearance. The chin, neck and lower belly are white and undertail-coverts yellow. The head is black from the lores over the crown to the nape and the ear-coverts. The iris varies from red to yellow; the legs and thinnish bill are black. Sumatran birds have darker upperparts.

Where to see: Most likely to be seen at forest edges, on Java and Sumatra.

Straw-headed Bulbul *Pycnonotus zeylanicus* 28–29cm

Indonesian name: Cucak Rawa

A melodious song is responsible for the popularity of this species as a cagebird in Indonesia. It is one of the largest bulbuls, recognised by its straw-coloured crown and ear-coverts and its black moustachial stripe. The back, wings and tail are shades of greenish-brown. It has a thin black eye-stripe, a white chin and throat, grey breast and belly, and yellow vent. The upper breast and back have a small amount of white streaking.

Where to see: Occurs on Java. Previously common in lowland and hill forest and riverine habitat, it is now Critically Endangered.

Sooty-headed Bulbul *Pycnonotus aurigaster* 19–21cm

Indonesian name: Cucak Kutilang

This bulbul has a short, almost vertical crest topping its lustrous black head from the chin and just below the eyes to the nape. Ear-coverts are whitish and underparts white tinged brown, darker on the upper breast and paler towards the yellow vent. Uppertail-coverts are white and tail black, tipped white. The hindneck is ash-grey darkening across the mantle and onto the back and becoming paler on the rump. The paler-edged feathers give a scalloped appearance. The wing-coverts and secondaries are similar. Primaries are brown, edged grey. Legs and bill black.

Where to see: Common in scrubby, open habitats on Java and Bali, often close to human habitation.

Orange-spotted Bulbul *Pycnonotus bimaculatus* 20cm

Indonesian name: Cucak Gunung

Plumage is darkish, brown to olive-brown above, with paler yellow-tinged ear-coverts and two distinguishing orange spots, one on the lores and the other above the eye. The throat is black shading into a brown upper breast that becomes off-white with brown mottling as it leads to a white belly and bright yellow vent. Some birds in central Sumatra have a greyish belly. The bill is black and legs grey to black. The iris is darkish red-brown. Juveniles are duller. The subspecies *P.b. tenggerensis* has a smaller orange eye spot and dark brown spotted ear-coverts.

Where to see: Occurs throughout the region, more often in lower mountain forests.

Sunda Yellow-vented Bulbul *Pycnonotus analis* 20–21cm

Indonesian name: Merbah Cerukcuk

As befits its name, has a bright yellow vent but is otherwise relatively sombre in plumage. The upperparts and tail are brown and the underparts white. The crown is dark brown and the supercilium white. Lores black. It has pinkish-grey feet and a black bill

Where to see: Common throughout the region, preferring more open habitat. Especially attracted to cultivation, plantations, gardens and parks, even in towns.

Melodious Bulbul *Alophoixus bres* 19–22cm

Indonesian name: Empuloh Jawa

This solidly built bulbul can be recognised by its habitual calling while raising its rufous crest and puffing out its white throat feathers. The upperparts from the head to the tail are greyish olive-brown with the uppertail tinged chestnut. The underparts vary from pale olive-brown on the upper breast to yellow and then buff on the vent with the flanks suffused olive-brown. The iris is brown, the legs are pink-brown and the bill is grey-brown.

Where to see: Occurs in the lower mountain forests of Java and Bali.

Arctic Warbler *Phylloscopus borealis* 13cm

Indonesian name: Cikrak Kutub

Identified by its conspicuous, long, yellowish-white supercilium above a blackish eye-stripe. It also has a single poorly defined white wing-bar. The upperparts are dark olive and the underparts almost white, blending to olive-brown on the flanks. Similar to the smaller Yellow-browed or Inornate Warbler (*Phylloscopus inornatus*), but much duller and with less obvious wing-bars.

Where to see: Occurs throughout the region as an overwintering migrant. Found generally at primary and secondary lowland forest edges and mangroves.

Sunda Warbler *Phylloscopus grammiceps* 10cm

Indonesian name: Cikrak Muda

This active little warbler frequently occurs in small groups in the lower canopy of forests in mountainous areas, foraging for insects. Mainly chestnut-and-grey upperparts, brown tail and white underparts. Long, black crown stripe extends to the mantle. It has two yellow wing-bars, a very pale grey eye-ring and orange legs.

Where to see: A montane forest species occurring throughout the region.

Mountain Leaf Warbler *Phylloscopus trivirgatus* 10–11cm

Indonesian name: Cikrak Daun

Mainly green above, with greenish-yellow underparts. Its prominent greenish-yellow median crown-stripe and supercilium are intersected by a blackish stripe along the crown side and a black eye-stripe. The legs are grey and the bill is black above with a reddish tinge on the lower mandible. Juveniles are duller and greener below.

Where to see: Occurs throughout the region. Common in montane forests, keeping to the treetops and forest edges.

Javan Tesia *Tesia superciliaris* 9cm

Indonesian name: Tesia Jawa

This almost tail-less little warbler creeps around on seemingly ungainly light brown legs. The upperparts from the nape to the tail are olive-green and underparts light grey, tinged olive on the flanks and rump. The crown and nape is black, as is the eye-stripe, which broadens at the nape. The long and broad supercilium and ear-coverts are grey-white. The bill is black above and yellow below.

Where to see: Occurs in West and Central Java in brushland and dense undergrowth at the edges and clearings of montane forest.

Yellow-bellied Warbler *Abroscopus superciliaris* 9cm

Indonesian name: Cikrak Bambu

The upperparts of this small warbler are olive-green becoming yellower on the rump and tail-coverts. The tail and wing feathers are brown, fringed olive-green. A long, whitish supercilium separates the dark grey crown, forehead and nape from dark grey lores and ear-coverts. Underparts are very pale grey on the chin, throat and upper belly and pale yellow below. The bill is brown, legs light pink and iris dark brown.

Where to see: Occurs on Java and Sumatra, where it adheres to forest undergrowth in the lowlands and foothills.

Pygmy Bushtit *Aegithalos exilis* 9cm
Indonesian name: Cerecet Jawa

A rather plain little tit with medium-long tail and a small bill. Upperparts are grey and brown and the throat light grey; the remainder of the underparts are buff, tinged pink. It has a very pale yellow iris, brown to black bill and yellow legs. Both sexes and juveniles are similar.

Where to see: Occurs only in the mountains of West and Central Java, showing a preference for montane forest edges, where it forages for insects, often in small flocks.

Javan Heleia *Apalopteron javanicum* 13cm
Indonesian name: Opior Jawa

This white-eye does not have the familiar broad white orbital ring. It has a grey crown extending onto the nape and the ear-coverts, and olive-green upperparts. The forehead is buff, as is the broken eye-ring, the iris is blackish-brown, the bill black and the legs dark olive-yellow. The throat and chest are buff blotched pale grey shading into a yellow, blotched grey belly and yellower flanks. The subspecies vary in the size and shape of the black loral patch.

Where to see: Occurs in thick undergrowth and forest of Java and Bali, sometimes venturing into cultivation.

Sangkar White-eye *Zosterops melanurus* 10–11cm

Indonesian name: Kacamata Biasa

An active bird with upperparts bright olive-green, and whole breast, belly, throat and vent bright yellow. Wings and tail are blackish-brown. It has a ring of bright white feathering around the eye broken by a dark loral stripe contrasting with darker sides to the face, and sometimes a yellowish forehead. The bill is dark brown and feet are grey.

Where to see: A locally common woodland species occurring on Java and Bali.

Warbling White-eye *Zosterops japonicus* 11–12cm

Indonesian name: Empuloh Jawa

Males have olive-green upperparts, yellowish on the rump. The forecrown is bright yellow at the lores becoming olive-green on the crown. It has a black loral spot and loral line passing under the white eye-ring. The throat and upper breast are pale yellow becoming pale grey to white on the underparts and buffish on the flanks. The undertail-coverts are pale yellow and bill and legs grey-black. Iris is pale grey.

Where to see: A mountaintop species occurring throughout the region.

Black-capped White-eye *Zosterops atricapilla* 9.5–10cm

Indonesian name: Kacamata Topi-hitam

The upperparts of this little white-eye are a solid olive-green becoming black on the forecrown, forehead, lores and a circle around the eye. The white eye-ring is broken by a black spot at the front. The blackish-brown flight and tail feathers are also edged olive-green. On the rump, the olive-green becomes yellower, and below, the bright yellow undertail-coverts extend to the lower breast in a wide stripe flanked by dark grey underparts. The throat and bib area are a lemon-green. The bill is fawn and black and legs grey.

Where to see: Occurs in the lower montane forests of Sumatra.

Lemon-bellied White-eye *Zosterops chloris* 11–12cm

Indonesian name: Kacamata Laut

This little white-eye has dullish green upperparts from the crown, over the back and wing-coverts to the yellow-green uppertail-coverts. The underparts are dullish pale yellow, brighter on the throat and undertail-coverts. There is a small yellow loral patch with the black lores extending below the unbroken white eye-ring to the greenish ear-coverts. The tail and flight feathers are blackish, edged green. The bill is blackish above and grey below and legs are grey.

Where to see: Occurs in extreme West Java and on Nusa Penida, Bali in a variety of habitats, from forest to mangrove and even cultivation.

Chestnut-capped Babbler *Timalia pileata* 15–17cm

Indonesian name: Tepus Gelagah

Distinguished by its masked appearance and longish tail. Main body colouration is warm olive-brown above and buffish-brown finely streaked black below. The head is more striking, with a bright chestnut crown underlined by a white forehead and supercilium. The chin, throat and neck are white, with the ear-coverts whitish, becoming pale grey towards the nape. The strong bill is black and iris and legs dark brown.

Where to see: This babbler inhabits scrubby and swampy areas, usually in the company of other babblers and occurs from West Java east to Bali.

Pin-striped Tit Babbler *Mixornis gularis* 11–12cm

Indonesian name: Ciung-air Melayu

The upperparts of this small tit babbler are all chestnut-brown, including the tail. The underparts, in particular the breast, are conspicuously but finely streaked with black on a ground colour of pale yellowish-green, becoming whitish on the lower belly and vent. Javan birds are more greyish below. The sides of the head are grey, sometimes tinged yellow, with a pale supercilium. The bill is brown and the legs are grey-blue.

Where to see: A common bird in lowland Sumatra, gathering in small groups foraging in dense thickets and bushes, especially bamboo, and secondary growth in broadleaf forest.

Golden Babbler *Cyanoderma chrysaeum* 10–12cm

Indonesian name: Tepus Emas

Golden Babbler has olive-green upperparts and a dull yellow, thinly streaked crown. The underparts are buff-yellow on the throat, becoming greyish on the flanks and buff on the vent. The tail and upperwing are grey tinged olive with flight feathers vaguely edged olive. Dark brown iris surrounded by dark grey-blue loral skin extending into a short black sub-moustachial stripe. The ear-coverts are yellow-grey, the bill is blackish above and black to pink below, and the legs are light brown.

Where to see: Occurs in montane forests of Sumatra.

Crescent-chested Babbler *Cyanoderma melanothorax* 13cm
Indonesian name: Tepus Pipi-perak

Upperparts are brown varying, across subspecies, from chestnut to dull olive-brown. Underparts are white to whitish-buff, becoming greyish or sandy-buff on the belly, with a black crescent shape across the upper breast. Ear-coverts are olive-grey to grey underlined by a black malar stripe. The supercilium is pale grey, the legs and bill brown.

Where to see: Occurs in the forests of Java and Bali.

Javan Scimitar Babbler *Pomatorhinus montanus* 21cm
Indonesian name: Cica-kopi Jawa

This bird has a striking greyish-black head with a long white eyebrow and a longish, decurved, buffish-yellow bill, black at the base. The throat and breast are white, the back, flanks and vent chestnut-brown, and the tail and wings dark brown. The feet are grey.

Where to see: Occurs on Java and Bali. Frequently seen in pairs or small groups with laughingthrushes, foraging in ground litter and through low bushes and in the lower tree canopy of higher montane forests.

Black Scimitar Babbler *Melanocichla lugubris* 25–27cm

Indonesian name: Cica-kopi Hitam

This all-black babbler has a substantial orange-red bill and a patch of bare blue skin behind the eye. Most of the plumage is dull black apart from shiny areas of the face, throat and crown, and bristly feathers of the lores and chin. The iris is dark brown (sometimes bluish) and the legs dark grey tinged brown. Juveniles are browner and duller.

Where to see: Occurs in the lower mountain forests of Sumatra.

White-bibbed Babbler *Stachyris thoracica* 18cm

Indonesian name: Tepus Leher-putih

Birds from West Java have all dark chestnut plumage, apart from the grey-black throat and face and a broad white breast-band. In West and East Java the crown to the nape is dark grey, with the white breast-band faintly edged blackish above and below. A small area of bluish bare skin surrounds a dark red iris; the bill is dark grey above, bluish below and legs are grey.

Where to see: Occurs in montane forest on Java. Subspecies *S. t. thoracica* occurs in West Java and *S. t. orientalis* in West and East Java.

Javan Black-capped Babbler
Pellorneum capistratum 16–17cm

Indonesian name: Pelanduk Topi-hitam

This brown-coloured babbler has prominent head markings. The crown to the extended nape feathers are black, contrasting with a pale orange-brown to buff supercilium, below which is a blurred broad grey-brown eye-stripe. The throat is white. The upperparts, upperwing and tail are brown and underparts pale orange-chestnut. The legs are brown, the bill grey and iris brown.

Where to see: Occurs in lowland forest on Java.

Abbott's Babbler *Malacocincla abbotti* 15–17cm

Indonesian name: Pelanduk Asia

Like Horsfield's Babbler, Abbott's is small and brown and has a heavy bill. It is distinguished by being lighter and duller and having a longer tail. The upperparts are brownish-olive, with a pale chestnut rump. The underparts are buffish-white on the throat, with a pale grey-green breast, buffish-brown belly and rufous undertail-coverts. The sides of the head are buffish-brown with a pale-grey supercilium.

Where to see: Locally common in Sumatra's lowland forests, with a few records for Java, but absent from Bali. Usually found at forest edges and in scrub and thickets.

Horsfield's Babbler *Malacocincla sepiaria* 15–16cm

Indonesian name: Pelanduk semak

This babbler has a dark grey-brown crown and warm russet-brown upperparts, becoming chestnut on the rump and tail. The flanks and vent are also chestnut, while the throat and belly are white and the breast grey. It has a heavy grey-and-black bill and red-brown iris. Separated from Abbott's Babbler by its darker and greyer crown. This noisy little babbler has a monotonous and repetitive call, *pee-o-eel pee-o-eet*, given particularly at dawn and dusk.

Where to see: This is a locally common resident throughout the region in dense undergrowth and low thickets of hill and submontane forests. Subspecies *M. s. sepiaria* occurs on Java and Bali and *M. s. barussana* on Sumatra.

Large Wren-Babbler *Turdinus macrodactylus* 19–21cm

Indonesian name: Berencet Besar

The upperparts of this large babbler are dark brown streaked black and light brown at the crown, becoming lighter and warmer over the upperwing-coverts and on the rump. The underparts are whitish flecked brown on the throat and grey-brown flecked whitish on the upper belly, forming a vague breast-band in Sumatran birds but missing in Javan. Javan birds are generally darker. A patch of bluish skin surrounds the eye. Ear-coverts dark brown. The bill is blackish and legs are dark brown.

Where to see: Occurs in lower primary forests and dense woodlands of Sumatra and Java.

Sunda Fulvetta *Alcippe brunneicauda* 14.5–15cm

Indonesian name: Wergan Coklat

This rather plain-coloured forest-dwelling babbler has a brown crown on a grey head. Mantle, wings and tail are brown, appearing more rufous on the wings, rump and uppertail-coverts; underparts are buffish-white. The bill is grey-brown, as are the legs and iris. Sexes are similar but juveniles appear paler with a yellow-brown bill.

Where to see: Occurs in Sumatra's lowland forests, where it prefers the tree canopy.

Javan Fulvetta *Alcippe pyrrhoptera* 14–15cm

Indonesian name: Wergan Jawa

This bird is dull chestnut-coloured from the crown over its back and wing-coverts, becoming warmer chestnut on the rump and tail. The crown is faintly striped with dark lateral stripes extending to the nape, bordered by a vague blackish-brown supercilium. The ear-coverts are light chestnut with fine white streaking. The throat and underparts are buff with browner flanks. The bill is blackish and legs are pale yellow-brown.

Where to see: Occurs in lower montane forests of West and Central Java.

Sunda Laughingthrush *Garrulax palliatus* 25cm

Indonesian name: Poksai Mantel

The head of this laughingthrush is dull grey down to the mantle, across the shoulders and to the lower belly; the belly becomes rufous-grey to the vent. The bristled lores, chin and face are black. The upperparts, including the wings, are rufescent brown-black; as is the tail. The iris is warm grey-brown surrounded by a large silver-blue patch of bare skin. The legs are dark brown and the bill is grey-black. Juveniles are duller and browner.

Where to see: Occurs in low montane forest on Sumatra. Now rare.

Spectacled Laughingthrush *Garrulax mitratus* 22–24cm

Indonesian name: Poksai Genting

This darkish grey laughingthrush has a chestnut crown and supercilium, a broad white eye-ring and bright orange-yellow bill and legs. It also has a chestnut vent and a white patch on the outer primaries. The bristly feathering to the lores, face and chin gives the impression of a black mask. A few feathers on the forecrown are whitish, the grey tail is tipped black and the iris is red-brown. Juveniles are duller and browner.

Where to see: A lower montane forest species occurring on Sumatra.

Long-tailed Sibia *Heterophasia picaoides* 28–35cm

Indonesian name: Sibia Ekor-panjang

A long-tailed grey babbler with white tips to its tail feathers and an obvious white wing patch. The head is grey and upperparts are tinged greyish-brown. Underparts are paler grey tinged brown on the belly. The vaguely downcurved bill is black, the iris is red-brown and legs are grey.

Where to see: Lives at the edges of broadleaved evergreen forest and lower montane forests on Sumatra, feeding on various insects, fruit, seeds and even flower buds.

Silver-eared Mesia *Leiothrix argentauris* 15–17cm

Indonesian name: Mesia Telinga-perak

The male is distinctively marked, with bright silver ear-coverts that contrast with a black crown and face, and a small yellow-orange tuft of feathers at the base of the bright yellow bill. The chin, neck and fore-collar are dark orange blending into an olive mantle and paler orange breast. The tail is blackish fringed yellow, the legs and iris yellow. Females are less bright.

Where to see: Occurs on Sumatra but is becoming rare in its preferred montane habitat because of trapping.

Velvet-fronted Nuthatch *Sitta frontalis* 12–13cm

Indonesian name: Munguk Beledu

Velvet-fronted Nuthatch has a red bill and reddish feet, dirty pinkish-white underparts and whitish throat. The upperparts are mainly violet-blue, with black-edged feathers in the tail and primaries, and the forehead is velvety-black. A very active and attractive bird which, unlike other nuthatches in the region, is well-adapted to creeping both up and down tree trunks in search of insects and spiders.

Where to see: Common in lowland and hill forests and plantations of Sumatra and Java. Usually seen in pairs but occasionally in small parties, moving erratically through the understorey.

Blue Nuthatch *Sitta azurea* 13cm

Indonesian name: Munguk Loreng

In shady forest it appears to have black
upperparts, but in fact its back, wings and
tail are shiny dark blue. Only the crown,
nape and sides of the face are black. The
throat and breast are buffish-white. East
Javan birds have a bluish-black lower
belly and vent, but these are black on
West Javan and Sumatran birds. It has
a yellow-grey bill and grey feet. This
nuthatch typically forages along branches
and tree trunks, levering off fragments of
bark to get at grubs and insects.

Where to see: Occurs in the Barisan
Range of Sumatra and on Java, where it
is common in lower montane forests.

Asian Glossy Starling *Aplonis panayensis* 20cm

Indonesian name: Perling Kumbang

Adults are all black and can be
separated from the Short-tailed
Starling (*Aplonis minor*) by being
larger and having a green iridescence.
It has dark red eyes and black legs.

On Bali they are smaller than the birds
in Java and Sumatra. Juveniles have a
yellowish iris, black-streaked whitish
underparts, and upperparts streaked
black and brown.

Where to see: A common
lowland bird occurring
throughout the region's
open areas near forest
and secondary vegetation
and coastal scrub, but
also attracted to coconut
plantations, cultivated areas
and gardens, and frequently
in towns and cities

Grosbeak Myna *Scissirostrum dubium* 20cm

Indonesian name: Jalak Tunggir-merah

This small starling is easily recognised by its predominantly dark grey plumage, the solid 'grosbeak-type' orange-yellow bill and the bright red, elongated uppertail-coverts, some with shiny tips. The legs are also orange-yellow and the iris is black-brown.

Where to see: Occurs on Java. Prefers tree canopy of lowland forest edges, where it often congregates in large flocks, feeding on fruit, seeds and insects.

Common Hill Myna *Gracula religiosa* 27–35cm

Indonesian name: Tiong Emas

Black overall with a purple sheen and orange wattles both behind and below the eye and has a white wing-bar. It has yellow feet and a stout orange bill. Another species under great pressure from habitat loss and trapping for the pet trade, this bird is particularly in demand for its ability to mimic other birds, as well as for its own enormous repertoire of whistles and calls.

Where to see: Although it occasionally congregates in small flocks, it is more likely to be seen in pairs in tall trees. Once a common bird of lowland forest edge, it has become rare on Bali and Java but remains in reasonable numbers on Sumatra.

Daurian Starling *Agropsar sturninus* 17–18cm
Indonesian name: Jalak Cina

The male is recognised by its grey head, a shiny purple patch on the rear crown and odd purple feathers on the upper mantle. The purple-black scapulars contrast with a broad white band formed by their tips. The rump is white-buff and the short tail black, glossed green. The underparts are whitish. It has a shiny purple lower mantle. The iris is brown, the bill is black and the legs are grey. Females and juveniles are paler and have more brown colouring.

Where to see: Occurs as a winter migrant to the region, often seen in city parks and gardens but also in coastal woodland.

Javan Pied Starling *Gracupica jalla* 22cm
Indonesian name: Jalak Suren

The upperparts are mainly black, tinged brown. It has a white wing-bar, with forehead, cheeks, belly, vent and rump also white. Bare orange skin around the eye, yellow feet and orange-white bill give the bird a colourful appearance. Agricultural practices and the pet trade have decimated the wild population, and even those seen in bird markets today may have been captive-bred.

Where to see: Occurs on Java and Bali. This species is now Critically Endangered, having once been common throughout the region, especially in more open and cultivated lowland country.

Bali Starling *Leucopsar rothschildi* 25cm

Indonesian name: Jalak Bali

Its distinctive plumage is mostly white, apart from black wing-tips and the tip of the tail. It has a long white crest, shorter in the female, and a patch of sky-blue skin around the eye. This endemic to Bali is one of the world's most threatened species, with very few birds now surviving in the wild. Destruction of habitat and trapping for the pet trade are the main causes of its decline.

Where to see: A small wild population has managed to survive under protection in its native range of north-west Bali (Bali Barat National Park). It has regrettably and unscientifically, recently been introduced to Nusa Penida Island, Bali, which is outside its native range.

Black-winged Myna *Acridotheres melanopterus* 23cm

Indonesian name: Jalak Putih

Apart from black wings and tail and a yellow patch of bare skin around the eye, this starling is all white. The tail is tipped white, and there is a white wing-bar. North-west Javan birds appear much paler and have black only on the flight feathers and tail. South-east Javan birds have a grey rump and shoulder. Birds from Bali have more contrasting plumage with white underparts, neck and head to top of mantle, and upperparts, including back and rump to the tail, grey-black.

Where to see: Three subspecies are endemic to the region: black-winged in north-west Java, grey-backed in south-east Java and grey-rumped on Bali. All seemingly prefer dry coastal grasslands.

A. m. tertius

Common Myna *Acridotheres tristis* 24–26cm

Indonesian name: Kerak Ungu

The upperparts, as well as the mid-belly and flanks, are dark warm brown and underparts dark grey at the neck, fading to brown and then white on the undertail-coverts. The head, neck and long crown feathers are shiny black contrasting with the strong yellow-brown bill and bright yellow bare skin below and behind the eye. When perched, a long white streak is visible on its forewing. The legs are yellow and the tail is brown, tipped white.

Where to see: Occurs on Sumatra and Bali in a wide variety of habitats, from cities to open country.

Sunda Thrush *Zoothera andromedae* 23–25cm

Indonesian name: Anis Hutan

This prettily patterned thrush has a particularly long bill. Its upperparts, including the crown, are grey-brown with black feather edging creating a scallop pattern. The sides of the face, including the ear-coverts and lores, are white stippled black, and there is a white eye-ring surrounding a black iris. The throat, breast and belly are pale grey, becoming white on the belly and undertail-coverts. The white flanks are beautifully patterned with bold black scallops. Legs are pinkish-grey and the bill is black. The sexes are very similar. Juveniles are browner above and buffish below.

Where to see: Occurs throughout the region in hill and mountain forests.

Horsfield's Thrush *Zoothera horsfieldi* 28cm

Indonesian name: Anis Horsfield

In typical thrush fashion, this species spends much of its time foraging on the ground. It has only a quiet whistling song and a feeble *tzeeet* alarm call. Its upperparts are olive-brown with golden-brown and blackish scaly markings, these being bolder on the back and upperwing-coverts. The brown-scaled underparts are more whitish, with a pale chestnut wash on the upper breast and flanks.

Where to see: Occurs throughout the region, on Java and Bali in dense wooded areas and montane forest. On Sumatra is confined to the montane woodland of the north.

Javan Cochoa   *Cochoa azurea*   23cm

Indonesian name: Ciung-mungkal Jawa

The male has crown and most of the upperparts and tail glossy blue, while the underparts are more purplish-blue. Bill and legs black. The female has a blue crown, wings and tail but is otherwise a very dark brown. Its high-pitched song is also quiet but when alarmed it emits a louder *djek-djek-djek* call, reminiscent of many other thrushes.

Where to see: Occurs in West and Central Java, frequenting the lower levels of tropical montane forest, seldom venturing into the canopy. It is often difficult to locate, as it seems to perch quietly and motionless for ages.

♀

♂

Siberian Thrush *Geokichla sibirica* 20–23cm

Indonesian name: Anis Siberia

The male is dark slate-blue above and slate-grey on the flanks, becoming whitish on the belly. The head and shoulder area appear a darker slate-blue, offsetting the ostentatious white supercilium. The undertail-coverts and outer tail feathers are tipped white. The legs are yellow and the bill black. Females are olive-brown and buff-brown below.

Where to see: Regularly overwinters on Sumatra and West Java, rarely on Bali. Occurs in the lower levels of moist forest, often close to water.

Chestnut-capped Thrush *Geokichla interpres* 16–17cm

Indonesian name: Anis Kembang

Chestnut-capped, as its name implies, this small thrush otherwise has a pied appearance. The chestnut crown extends onto the nape, stopping at the slate-grey mantle and upperparts. It has two distinctive broad white wing-bars. The throat and upper breast are dense black, leading into a black-spotted lower breast and flanks and white belly and vent. The bill is blackish-brown and the legs pink.

Where to see: Occurs low down in primary, secondary and even logged forest on Java, rarely on Bali and Sumatra.

Orange-headed Thrush *Geokichla citrina* 20–22cm

Indonesian name: Anis Merah

The usual haunts of this thrush are more secluded forest among thick undergrowth, but its high-pitched whistling alarm call betrays its presence. It has a very nice song and this, along with its attractive colouring, makes it a prime target for local bird-keepers and, more recently, the export pet trade. Its head, nape and belly are bright orange, the back grey-blue, and the tail and wings dark grey. It has a white vent and wing-bar. The female has greenish-brown upperparts.

Where to see: Occurs in Java and Bali, in lowland and hill forest and bamboo thickets. Uncommon.

G. c. rubecula ♂

Island Thrush *Turdus poliocephalus* 18–22cm

Indonesian name: Anis Gunung

This is a medium-sized thrush with a dark grey head blending into dull blackish upperparts. The throat and upper breast are mid-grey, while the belly is chestnut and the vent white. The eye-ring, bill and legs are yellow. It has a pleasantly melodious song and a rather raucous rattling alarm call.

Where to see: Occurs in the highest mountains of Sumatra and Java. More common on Java, where it can usually be seen close to the volcano of Mt. Tangkuban Perahu and in the mossy forests of Mt. Gede. On Sumatra, the peaks of Mt. Leuser are its home.

Asian Brown Flycatcher *Muscicapa dauurica* 12–14cm

Indonesian name: Sikatan Bubik

This species of submontane forest and forest edges can frequently be found in plantations, open forest and even gardens. It hunts aerial insects from exposed perches, often shaking its tail on arrival back at the perch. It is greyish-brown above and whitish below, becoming grey-brown on the flanks and the sides of the breast, and has a pale eye-ring. The black bill has a yellow base to the lower mandible.

Where to see: A winter migrant to the region from north-east Asia, but some may be resident in north Sumatra. Occurs in woodland, forest and scrubland, sometimes close to cultivation.

White-rumped Shama *Copsychus malabaricus* 21–27cm

Indonesian name: Kucica Hutan

The male has a very long tail, about half its total length, with dull black central feathers and white outers, which it habitually flicks on landing. Its head, neck, back and wings are black with a blue sheen, its lower underparts are deep orange, and it has a conspicuous white patch on the rump. Another beautiful songster in decline, due to trapping for the pet trade.

Where to see: Occurs on Sumatra and Java, spending a lot of time foraging on the ground in low dense undergrowth. Common on Sumatra in denser forest, but far less so on Java.

Oriental Magpie-robin *Copsychus saularis* 20cm
Indonesian name: Kucica Kampung

Sumatran birds have black wings and central tail, with white outer tail feathers, belly and vent, and a broad white patch across the wing-coverts, whereas those from East Java and Bali are all black except for the white wing-covert patch. Females are much duller, with the black plumage replaced by grey. Juveniles have similar but mottled plumage.

Where to see: Occurs throughout the region in lowland mangroves and forests to town gardens. Once common but trapping for the pet trade has caused a serious decline.

C. s. pluto ♂

C. s. musicus ♂

Pale Blue Jungle-flycatcher *Cyornis unicolor* 16–18cm

Indonesian name: Sikatan-rimba Biru-muda

Males of this mid-sized, longish-tailed flycatcher are predominantly blue, whereas females are grey-brown. The male has black lores and a pale-blue forehead grading into a short supercilium. The throat is pale blue grading to grey on the belly and white on the lower belly and undertail-coverts. The female has grey-brown upperparts and buffish underparts.

Where to see: Occurs in the lower dense and moist primary forests of Sumatra and Java.

Javan Jungle-flycatcher *Cyornis banyumas* 14–15cm

Indonesian name: Sikatan-rimba Jawa

Males have dark blue upperparts with iridescence on the forehead, eyebrow and shoulder. The chin, lores and a thin line around the eye are black, while the throat and breast are rufous-orange, grading into white on the lower belly and vent. The bill is black fading to dull purple at the base, the legs vary from dark brown to pale pink and the iris is dark brown. Females have brown upperparts with an off-white eye-ring, and underparts similar to the male's but paler.

Where to see: Occurs on Java, where it favours low undergrowth and bamboo thickets and the edges of glades in high montane forests.

Mangrove Jungle-flycatcher *Cyornis rufigastra* 15cm

Indonesian name: Sikatan-rimba Bakau

Characteristic blue upperparts and orange-red underparts. The bill is black and the legs blue-grey, although the female is paler and has a whitish chin and a white loral patch. The many subspecies vary, typically in strength of colour of upperparts and underparts. It is most likely to be confused with the Hill Blue Flycatcher (*Cyornis banyumas*), but that has a pale blue forehead and blackish chin.

Where to see: Occurs in mangrove forest on Sumatra and West to Central Java, also in Baluran National Park in East Java and Bali Barat National Park in Bali.

Sumatran Niltava *Niltava sumatrana* 15cm

Indonesian name: Nilatava Sumatera

The male's dark blue upperparts are iridescent on the forecrown, rump and uppertail coverts and tail and dark blue over the back and wing-coverts. The flight feathers are blue-black, edged dark blue. There is a thin iridescent blue neckline and patch on the shoulder. The lores and forecrown, chin, throat and ear-coverts are blue-black. The underparts from the neck to the undertail-coverts are bright orange. The legs are blue-grey and the bill black. Females are grey from crown to nape, the remainder of the upperparts and sides of the face being warm brown. The underparts are grey at the belly, becoming whitish at the vent.

Where to see: Occurs in montane forest in the Barisan Range of W. Sumatra.

Blue-and-white Flycatcher *Cyanoptila cyanomelana* 16–17cm

Indonesian name: Sikatan Biru-putih

The male has mainly rich blue upperparts, over the crown, the upperwing-coverts and along the edges of the tail and flight feathers; the greater part of the flight feathers being black. The lores, face, ear-coverts, chin and upper breast are shiny black. The remainder of the underparts, including the undertail-coverts, are white. The legs are purplish-black. The bill is black and the iris dark brown.

The female's upperparts, including the crown, are grey-brown, the ear-coverts brown finely streaked buff and the wings dark brown with russet edges. Tail feathers are also russet-brown. The underparts are buff with a warmer tint to the breast.

Where to see: Overwinters in the hill and montane forests of Sumatra and Java.

♂

Indigo Warbling-flycatcher *Eumyias indigo* 14cm

Indonesian name: Sikatan-kicau Ninon

This species is mainly a deep indigo-blue, especially on the upperparts and breast, the blue becoming dense blue-black on the throat and face towards the bill base. The forehead is bluish-white, as is the supercilium. The belly is grey-blue, becoming white and then buff on the vent. The legs and bill are black. Sexes are very similar but juveniles have dull, lightly barred upperparts and buff-spotted underparts.

Where to see: A relatively common resident of montane and submontane forests on Java, but less so on Sumatra.

Verditer Warbling-flycatcher *Eumyias thalassinus* 15–17cm

Indonesian name: Sikatan-kicau Hijau-laut

This is a large blue flycatcher with an upright stance. The male's plumage is a fairly uniform turquoise-blue, with brighter forehead and throat and white fringing to the undertail-coverts giving a scaled appearance. The dark grey area on the face does not reach the black lores. The feet and bill are blackish and iris is dark brown. Females look duller and greener and juveniles are a mottled brown tinged with green.

Where to see: Occurs on Sumatra, where it prefers open forest edges, choosing exposed perches from which it can launch its attacks on passing airborne insects.

Javan Whistling-thrush *Myophonus glaucinus* 25–26cm

Indonesian name: Ciung-batu Jawa

The male has blackish underparts and dark blue upperparts with an often hidden, brighter blue area on the shoulder. The plumage is more uniform than that of other whistling thrushes, lacking spangles. It has a blackish bill and legs. The female is browner but duller, with juveniles a bit darker and marked with elongate whitish spots.

Where to see: Occurs on Java and Bali, keeping to the lower levels in montane forest. Often secretively hides away in dark corners and crevices but is occasionally seen foraging on the ground.

Blue Whistling-thrush *Myophonus caeruleus* 29–35cm

Indonesian name: Ciung-batu Siul

A largish blue-black thrush having a moderate purplish sheen and sparsely flecked and spangled white on the wing-coverts. The contrasting yellow bill is often marked with black, particularly on Sumatran birds. Legs are black. It has a high-pitched screeching alarm, *screech-chit-chit-chit*, and a loud whistling song often incorporating imitations of other birds' songs. A ground feeder, taking insects, invertebrates and small fruits.

Where to see: Uncommon on Java and Sumatra but absent from Bali. Generally found in denser lowland and hill forests beside rivers and among exposed limestone rock formations.

Javan Forktail *Enicurus leschenaulti* 28cm

Indonesian name: Meninting Jawa

The deeply forked tail consists of tiers of white-tipped black feathers overlaid to form an attractive, evenly spaced, crescent-shaped pattern pointing up to the body. The outer tail feathers are all white. The head is all black except for a contrasting white upstanding crown. The chin, breast and back are also black, as are the shoulder and scapulars. It has an obvious white wing-bar and white rump and belly. The legs are pink and the bill black. Males and females are similar, with juveniles being browner and duller.

Where to see: Occurs in the lowland forests of Java and Bali.

Yellow-rumped Flycatcher *Ficedula zanthopygia* 13–14cm

Indonesian name: Sikatan Emas

The underparts and lower back and rump of this flycatcher are bright yellow, and the undertail-coverts whitish. The upperparts are all black apart from the broad white supercilium and a white area on the wing-coverts and tertials. The bill is black, grey-blue below; legs black. Females are olive-grey over the crown to the yellow uppertail-coverts. Underparts whitish on the chin becoming pale yellow on the breast and whitish again on the vent. The pale wing patch is much less obvious than in the male.

Where to see: A migrant that occurs throughout the region and is frequently found at woodland edges.

♂

Mugimaki Flycatcher *Ficedula mugimaki* 12.5–13.5cm

Indonesian name: Sikatan Mugimaki

The male has blackish-grey upperparts including the head and tail, with two prominent white spots, one just behind the eye, the other on the tips of the median coverts. The belly and undertail-coverts are also white. The remainder of the underparts up to the throat are orange-red. The small bill is black, the legs are pink and the iris is dark brown. The female has dark brown upperparts and tail, the throat to mid-belly pale orange-red and the belly white.

Where to see: Occurs throughout the region, mainly in lowland broadleaf evergreen forest.

♂

♂

Snowy-browed Flycatcher *Ficedula hyperythra* 11–13cm

Indonesian name: Sikatan Bodoh

This delicate little flycatcher has a roundish head and short tail. The underparts are alike in both sexes, being pale chestnut on the chin blending to fawn on the belly and white below the tail. The upperparts differ, the male being dark steely blue above and the female olive-brown. The male has a noticeable white stripe above black lores almost meeting centrally but extending over the eye. The leg colour is variable, from purple to grey in males to pink in females. The bill is black and iris very dark brown. Females are coloured orangey-buff from above the bill to just over the eye. Male Sumatran birds have blackish wings and tail, and those on Java and Bali are greyer above.

Where to see: Common in the highland forests of west Sumatra and on Java and Bali

Little Pied Flycatcher *Ficedula westermanni* 11cm

Indonesian name: Sikatan Belang

This species is sexually dimorphic, the male being black and white and the female brown and white. The male's upperparts are all black, apart from a white wing-bar and supercilium and white edges to the base of the outer tail feathers. The underparts are all white and the legs and bill black. Iris dark brown. Females have brown upperparts and whitish underparts, the tail being rufous-brown. Juveniles are heavily spotted with buff on the head and upperparts and have dark barring on the breast. It has a quiet *pi-pi-pi-pi* call followed by a vibrating churring sound.

Where to see: Occurs throughout the region in submontane to montane forest.

Rufous-chested Flycatcher *Ficedula dumetoria* 11–12cm

Indonesian name: Sikatan Dada-merah

The male's upperparts are mainly black from the head and mantle, including the upperwing and tail. The long supercilium is white as are the median covert tips and some of the upperwing-coverts, which form a white stripe along the wing. The bases of the outer tail feathers are also white. The throat and upper breast are orange-red, the remaining underparts white. Females are similar to the male but coloured olive-brown instead of black above.

Where to see: Occurs throughout the region, but rarely on Bali. It is mostly insectivorous, preferring mainly lowland forests.

Pied Bushchat *Saxicola caprata* 13–14cm

Indonesian name: Decu Belang

Males are all black, apart from a white rump, vent and distinctive wing-bar and shoulder patch. Females have dark brown-streaked upperparts, dark brown tail, buff underparts with rufous flanks, and pale rufous-brown rump. Juveniles are spotted dark and light brown. Often sings from a prominent perch while cocking its tail.

Where to see: Like all the chats, this bird prefers scrubby grassland, where it perches on exposed bushes or posts to locate its insect prey. A common resident of the dry, open lowlands of Java and Bali, occurring sporadically in the hills.

Greater Green Leafbird *Chloropsis sonnerati* 18–22cm

Indonesian name: Cica-daun Besar

Of the many leafbirds resident in the region, the Greater Green Leafbird is by far the largest. It is bright green on the back and tail, with slightly paler underparts. The strong hooked bill is black. The male has a black throat and face with a blue malar stripe, and the shoulder of the wing is marked by a small blue spot. The female has a yellow throat edged by a blue malar stripe, with a distinctive yellow eye-ring. Juveniles are very like the female, with the blue malar stripe being very dull or non-existent.

Where to see: Typically found with other leafbirds in lowland primary or secondary forest throughout the region.

Blue-winged Leafbird *Chloropsis moluccensis* 17cm

Indonesian name: Cica-daun Sayap-biru

Both subspecies have the typical colour of most leafbirds; generally yellowish-green above and more yellow below. This species is recognised by the blue wings and tail. The male has a black patch below the eye which extends into a yellow-bordered black throat.

The Sumatran subspecies has a yellow crown and yellow to orange nape, whereas the Javan subspecies has a green crown and yellow upper breast. Females lack the black head and throat markings. All have a blue malar stripe, a little less evident on females.

Where to see: Occurs at forest edges on Sumatra and Java.

Asian Fairy-bluebird *Irena puella* 21–25cm

Indonesian name: Kecembang Gadung

Apart from the black face, neck, underparts, tail and primaries, the male's plumage is brilliant blue. The iris is red and legs are dark grey. Females are a duller blue with a greenish tinge, except for their brighter blue rump. It can be distinguished from other bright blue-and-black birds by its larger size.

Where to see: Commonly found in the lowland forests of Sumatra, but less so on Java. More likely to be seen at fruiting fig trees, where it is often in mixed flocks with other birds. Otherwise it confines itself to foraging through the tops of tall trees of primary and secondary forest.

Crimson-breasted Flowerpecker *Prionochilus percussus* 10cm

Indonesian name: Pentis Pelangi

The adult male is brilliantly coloured, with bright yellow underparts broken only by a bright crimson breast patch. The upperparts are greyish-blue, with a bright red patch on the crown. The forehead and primaries are black and the tail is dark blue-grey. It has a white malar stripe finely underlined in black. Females are olive-green above, with only a tinge of orange on the crown, and have a yellow-tinged grey throat, a white malar stripe and a yellow belly with greyish-olive flanks; the undertail-coverts are distinctively white.

Where to see: A rather scarce and local bird of Sumatran and West Javan lowlands.

Modest Flowerpecker *Pachyglossa modesta* 9–10cm

Indonesian name: Cabai Gesit

This flowerpecker is comparatively dull, appearing as a small, dingy, grey-brown bird that continuously twitches its tail from side to side, with an unusually thick bill for a flowerpecker. The upperparts are shaded olive-green and underparts are buff with indistinct brown streaks. A buff submoustachial stripe separates the light brown malar stripe from the pale brown ear-coverts. The brown tail and primaries are edged paler greenish-brown. The iris is bright red-brown, the legs brown and the bill blue-grey. Sexes are similar, while juveniles are plainer and yellower below.

Where to see: Occurs in woodlands in the lowlands of Sumatra and Java.

Orange-bellied Flowerpecker *Dicaeum trigonostigma* 8–9cm

Indonesian name: Cabai Bunga-api

The orange belly, vent, rump and lower back, and the blue-grey head, wings and tail are diagnostic of the male. The female is pale olive-grey with a grey throat, becoming yellowish on the belly, the rump being dirty orange. The pointed and decurved bill is black and the legs grey. Juveniles look like dull females, lacking yellow or orange.

Where to see: Occurs throughout the region's lowlands, flitting among the forest canopy, garden trees and mangroves. It is very active, searching out small insects as well as small ripe fruits.

Javan Flowerpecker *Dicaeum sanguinolentum* 9–10cm

Indonesian name: Cabai Gunung

The male has a blood-red chest, becoming buff on the throat. The belly is pinky-buff down to the undertail-coverts. A broad black stripe is superimposed down the centre of the breast to the belly. The upperparts to the top of the head are lustrous steel-blue. The bill and legs are black and iris is blue-brown. Females have a bright red rump on an otherwise olive-brown body, tinged orangey-buff on the underparts. Juveniles are similar to females but show more grey on the head and a yellow tinge to the underparts.

Where to see: Occurs mainly in montane and hill forest of Java and Bali but is occasionally found in south Sumatra.

Scarlet-headed Flowerpecker *Dicaeum trochileum* 8–9cm

Indonesian name: Cabai Jawa

Adult males are scarlet-red from the head to the lower breast and along the back to the uppertail-coverts. The tail and upperwings are black, with a very pale grey belly becoming darker on the flanks and lower breast. The feet and bill are black and the iris brown. Females have light brown upperparts and a scarlet-red rump. The head and mantle often appear reddish-brown. The underparts are pale grey-white. Juveniles are light green-brown on the upperparts but show orange on the rump.

Where to see: This is a coastal and lowland species, feeding on insects, nectar and fruit and occurring throughout the region.

Ruby-cheeked Sunbird *Chalcoparia singalensis* 10–11cm

Indonesian name: Burung-madu Belukar

Although this small sunbird is brightly coloured, when in the shade its iridescence is not often visible, and males even appear blackish. In bright light, however, the male's iridescent dark green upperparts and crown are brilliant, and the ear-coverts are a deep red. The throat is orange-brown, grading into a yellow belly. Females are duller, with olive-green upperparts and pale orange-and-yellow underparts. The straight bill and legs are black, the legs often with a greenish tinge.

Where to see: This is a bird of forest edge and sparse woodland, often found in the company of other species. Occurs on Sumatra and Java.

Brown-throated Sunbird *Anthreptes malacensis* 12–13cm

Indonesian name: Burung-madu Kelapa

Identified by its bright colouration and long, curved bill, this species usually hovers in front of flowers in search of nectar. The male has a yellow breast, belly and vent, and a brown throat often fringed with dark purple. The face and chin are olive and the upperparts are iridescent olive-green, blending into a dark bluish head with a green sheen. The female is olive-green above and yellowish below. Of the region's 12 different sunbirds, only Brown-throated and Ornate occur on Bali.

Where to see: Occurs throughout the region. Usually seen in open plantations, gardens, coastal scrub and mangroves.

Van Hasselt's Sunbird *Leptocoma brasiliana* 9–10cm

Indonesian name: Burung-madu Pengantin

Males have primarily dark blackish upperparts, an iridescent emerald-green crown and nape and blackish-blue tail. The lower mantle, neck and sides to the face are solid black. Some green iridescence is present on the scapulars and uppertail-coverts. The throat and belly are maroon and belly dull purple. The female is dark olive above and paler olive below.

Where to see: Occurs on Sumatra and Java, mainly in lowland forest, feeding on insects, nectar, fruit and seeds.

♂

Copper-throated Sunbird *Leptocoma calcostetha* 12–13cm

Indonesian name: Burung-madu Bakau

Males are dark, even appearing blackish. Upperparts have green-purple iridescence and the breast and malar stripe purple. The throat and upper breast show a dark coppery-coloured sheen, while the flanks are yellow and the tail bluish-black. Separated from the smaller Purple-throated Sunbird (*Leptocoma sperata*) by lacking the red breast but having yellow flanks. Females have a greyish head, dark olive-green to brown upperparts and black tail. Undertail-coverts and throat are pale grey, grading into a greenish-yellow breast.

Where to see: Occurs along the coastal lowlands of Java and Sumatra and found mostly in coastal scrub, woodland and mangroves.

♂

Ornate Sunbird *Cinnyris ornatus* 10cm

Indonesian name: Burung-madu Sriganti

Quite small, with a long, curved black bill, it has olive-green upperparts with darker wings and a black tail. The underparts are bright yellow, with white showing below the tail. Breeding males have a black chin and upper breast with a purple iridescence. A very active and often noisy little bird, either flitting between flowers or hovering to extract nectar. It also feeds partly on insects and pollen. The nest, a wonderfully woven structure of fine grasses and hair-like materials, is suspended precariously from the end of a branch or among foliage.

Where to see: Occurs throughout the region, particularly in lowland areas of scrub, mangroves and woodland.

♂

Javan Sunbird *Aethopyga mystacalis* 9–12cm

Indonesian name: Burung-madu Jawa

The male is scarlet over the head, back and breast, with forecrown and malar stripe iridescent purple. The wings are olive-grey with the upperwing-coverts dark crimson to blackish. The underparts are bright white merging into pale yellow on the vent. The rump is also yellow. The long tail is iridescent dark purple. The iris is blackish-brown and bill and legs are brown. The female has olive-green upperparts and a grey head with pale yellowish breast and whitish belly. Juveniles are similar but greyer.

Where to see: Occurs in submontane forested regions of Java.

♂

♂

Crimson Sunbird *Aethopyga siparaja* 10–15cm

Indonesian name: Burung-madu Sepah-raja

The male has a blackish forecrown with an iridescent purple-green sheen. From the back of the head, over the mantle and to the upperwing-coverts, including the chin to mid-breast, the plumage is crimson. The lower breast to vent is blackish-purple. There is an iridescent purple-blue moustachial stripe. The rump is yellow and remaining upperparts dark brown to black. The smaller females are olive-green above and have light olive-buff underparts.

Where to see: Occurs in more open forest and forest edges of south Sumatra and Java, occasionally close to human habitation.

♂

White-flanked Sunbird   *Aethopyga eximia*   13cm

Indonesian name: Burung-madu Gunung

This colourful sunbird is aptly named after the cluster of soft white feathers on its flanks. Apart from the rather drab olive-and-black wings, longish blue-green tail and olive underparts, the male has a bright yellow rump and red throat and upper breast, the last crossed by a bluish-green necklace-like band, and his crown is iridescent purple. Except for the white flanks, the shorter-tailed female is dull olive, a little paler on the throat and vent.

Where to see: Endemic to Java, occurring in mountain forests and alpine scrub. Particularly fond of flowering trees.

Little Spiderhunter   *Arachnothera longirostra*   13–16cm

Indonesian name: Pijantung Kecil

An inconspicuous little olive-and-yellow bird with a long, curved bill, which is black above and grey below. The upperparts are drab olive-green and the underparts brilliant yellow. It can be separated from all similar species by its whitish throat and black moustachial stripe.

Where to see: Common in lowland and hill forests throughout the region, and in some areas deep into the mountains. It inhabits gardens, plantations and logged forests, where it can often be seen moving rapidly across open areas as it searches for nectar, in particular from the flowers of banana and ginger.

Long-billed Spiderhunter *Arachnothera robusta* 21–22cm

Indonesian name: Pijantung Besar

This long-billed and largish spiderhunter is normally a solitary bird, even to the extent of driving away all other spiderhunters that encroach on its territory. It has a habit of perching high on exposed branches and uttering its repeated and uninteresting call, *chew-lewt, chew-lewt*. In flight, it gives a simple *chit-chit* call. Its upperparts are olive and its underparts yellow, with the breast and throat streaked olive-green. Separated from the Spectacled Spiderhunter (*Arachnothera flavigaster*) and the Yellow-eared Spiderhunter (*Arachnothera chrysogenys*) by lacking yellow ear patches or an eye-ring.

Where to see: Occurs at forest edges on Sumatra and Java.

Eurasian Tree Sparrow *Passer montanus* 14–15cm

Indonesian name: Burung-gereja Erasia

This is perhaps the commonest of all resident species throughout the region. It has a chestnut crown, white face, black ear patches, pale buff throat, sides of face and underparts, brown back, wings mottled with black and white, and two pale wing-bars. The bill is bluish-black and legs are brown. Males and females appear alike, with juveniles looking paler.

Where to see: Well distributed throughout the region, particularly in the lowlands, where it has colonised almost all areas, from cities to forest clearings. Commonly seen feeding on the ground, and in large flocks raiding seeding crops.

Streaked Weaver *Ploceus manyar* 15cm

Indonesian name: Manyar Jambul

Breeding males of this colonial species have a black head with golden-yellow cap, dark brown upperparts with russet-brown feather edges giving a streaked appearance, and white underparts becoming brownish on the flanks and black streaking on the breast. It has a black bill and pink legs. Females and non-breeding males have a black-streaked brown head and buff eyebrow and chin. Large nesting colonies are established in isolated trees near good feeding areas, the polygamous males leaving the females alone to weave their elaborate suspended nests.

Where to see: Occurs on Bali and Java, more often in large roaming flocks, stopping off at seeding grassland and often around rice fields and reedbeds.

Baya Weaver *Ploceus philippinus* 15cm

Indonesian name: Manyar Tempua

The breeding male has a bright yellow crown and nape and blackish-brown cheeks and throat, the upperparts being streaked and mottled dark and light rufous-brown and the underparts a warm buff. Females have the crown streaked and mottled brown, a whitish chin, a buff eyebrow, and light brown cheeks and upper breast. It has a black bill, brown wings and tail, and a brown iris. In females and non-breeding males the forehead, crown and nape to the back become brown with darker streaking. Juveniles are much like females but lack heavy streaking or the obvious buff supercilium.

Where to see: Occurs throughout the region, usually in grassland and overgrown vegetation and often close to human habitation.

Asian Golden Weaver *Ploceus hypoxanthus* 15cm
Indonesian name: Manyar Emas

The spectacular male in breeding plumage has a black face and throat and bright golden-yellow head and body all the way to the rump. The feathering of the upperwings and tail is brown with buff-and-yellow borders, and the mantle is streaked brown. It has pink legs, a black bill and blackish iris. Female and non-breeding males are similar, with dark brown, buff-streaked upperparts, including the forehead and crown. The wings and tail are dark brown. The broad supercilium and cheeks are orangey-buff. Underparts are pale orange-buff grading towards white undertail-coverts. Her bill is pink as are the legs.

Where to see: Occurs in the lowlands of Sumatra and West Java, often close to water and near cultivation, where it relies mainly on a seed diet.

Red Avadavat *Amandava amandava* 9.5–10cm

Indonesian name: Pipit Benggala

For most of the year males and females look alike, but as the breeding season approaches, sexually dimorphic plumage becomes very apparent. Females have greyish-brown upperparts and greyer sides to the head, with greyish-buff underparts and blackish wings and tail. The rump and bill are bright red. Breeding males become crimson-red on the back and breast, with small white spots on breast sides and flanks and also on the rump. The feet are pinkish-grey. It feeds mainly on seeding grasses but also occasionally insects.

Where to see: The red-bellied subspecies occurs on Java and Bali and inhabits grassland and marshy areas often close to cultivation.

♂ ♀

♂

Sunda Zebra Finch *Taeniopygia guttata* 10cm
Indonesian name: Pipit Zebra

The male has grey upperparts from the top of the head to the white rump, where the uppertail-coverts are barred with black. The ear-coverts are orange to dark orange and there is a grey eye-ring. From the lores a white, teardrop-shaped area thinly bordered black stands out from the grey area below the chin to the upper breast. There is a narrow black breast-band, below which the underparts become buffish-cream. The tail is black, iris dark orange and the bill red-orange. Legs pink. Females have much paler ear-coverts.

Where to see: Occurs throughout the Lesser Sundas, from Lombok eastwards. Occasionally appears in the more open, lowland coastal areas of Bali.

Tawny-breasted Parrotfinch *Erythrura hyperythra* 10cm
Indonesian name: Bondol-hijau Dada-merah

The upperparts from the top of the head to the rump are green becoming orange-brown on the uppertail-coverts. The forehead is black then blue where it meets the green on the top of the head. The underparts and face are a warm buff, the legs pinkish-red, the bill black and the iris brown. Females are duller and juveniles paler without blue on the forehead.

Where to see: Occurs on Java and occasionally on Bali. It is mainly seen in moist montane forests and bamboo thickets, although it also occurs in open grassland and rice fields.

Pin-tailed Parrotfinch *Erythrura prasina* 12–15cm

Indonesian name: Bondol-hijau Binglis

This species has green upperparts, the rump and long, pointed tail are red, and the lower belly and vent brownish-buff. The male has a red patch in the centre of the belly and blue throat, cheeks and forehead. It has a black bill becoming grey at the base, pink legs, dark brown iris and a dark grey eye-ring. A rare colour variation has the red areas of plumage replaced by yellow. The female has a shorter tail, buff underparts, a greenish head, and a dull red tail, rump and uppertail-coverts. Juveniles appear paler than the female.

Where to see: Occurs on Sumatra and Java. Inhabits both montane and lowland moist forest, but also found in scrubland, bamboo thickets and rice plantations, often mixing with large flocks of munias.

red-morph ♂

♀
♂
golden morph

Javan Munia   *Lonchura leucogastroides*   11cm
Indonesian name: Bondol Jawa

This munia has dark chestnut-brown upperparts, crown, tail and vent; black chin, throat and sides of face down to the upper breast; and a white belly. It should not be confused with the White-bellied Munia (*Lonchura leucogastra*), which has fine pale streaking on the upperparts, brown flanks and a yellowish-brown tail. The bill is pale greyish to blue. The legs are grey and the iris dark brown. Females are similar, and the juvenile is buff-brown with whitish underparts, spotted brown on the flanks and lower breast.

Where to see: Occurs in south Sumatra, Java and Bali. Normally a grassland bird, it is attracted with other munias to the easy pickings of cultivated land and especially rice fields at harvest time.

Black-faced Munia   *Lonchura molucca*   10–11cm
Indonesian name: Bondol Taruk

This munia has a black crown and ear-coverts and chin to the upper breast. The nape to the blackish upperwing-coverts is brown. The tail and uppertail-coverts are black. The lower breast and below, to the undertail-coverts, are white lined with narrow blackish barring. The bill is grey-blue and legs are grey. A grey eye-ring surrounds a black iris.

Where to see: Occurs sporadically on Java and Bali. A typical munia that feeds on seeding grasses, often in rice fields.

Scaly-breasted Munia *Lonchura punctulata* 10–12cm

Indonesian name: Bondol Peking

This species is easily identified by the scaly appearance of its breast and flanks (that show as double scales in this subspecies). White feathers edged with brown. The upperparts and throat are brown without streaking. It has a red-brown face and throat. Bill is dark grey above, paler below, legs are grey and the iris is dark reddish-brown. Juveniles are paler and have a black bill.

Where to see: Occurs throughout the region. It inhabits scrub, gardens and open grassland but is considered a pest to the farmer for raiding rice fields and crops.

Black-headed Munia *Lonchura atricapilla* 11–12cm
Indonesian name: Bondol Rawa

Apart from its black head, neck and chin to upper belly, the plumage is dark chestnut with a black-brown patch on the centre of the lower belly as in the north Sumatra subspecies (illustrated). The bill is pale blue-grey and legs dark grey. A dark grey eye-ring surrounds a brown iris.

Where to see: Occurs on Sumatra, often in large flocks feeding from seeding grasses, but also in rice fields.

White-capped Munia *Lonchura ferruginosa* 11–12cm
Indonesian name: Bondol Oto-hitam

This munia has a whitish head inclusive of the nape and ear-coverts and the chin to upper belly are black. The remaining plumage is dark chestnut with a black-brown patch on the centre of the lower belly reaching the undertail-coverts. The bill is pale grey and legs medium grey. A dark grey eye-ring surrounds a brown iris.

Where to see: Occurs on Java and Bali and very occasionally in south Sumatra. Often seen in large flocks feeding on seeding grass heads. Also in rice fields.

White-headed Munia *Lonchura maja* 11cm

Indonesian name: Bondol Haji

Quite small, with the head and throat completely white and a greyish nape. The remainder of the plumage is chestnut-brown, apart from a small area of black on the belly. The bill and feet are pale bluish-grey and the iris is dark brown. Males and females are similar, the females often showing a darker crown and nape and paler underparts. Juveniles are brownish above and buff below with a creamy-white chin and throat.

Where to see: A widespread lowland bird occurring throughout the region. Common on Java and Bali, where it prefers open grassland, as well as swamps, reedbeds and rice fields.

Pale-headed Munia *Lonchura pallida* 11cm

Indonesian name: Bondol Kepala-pucat

A small munia with a whitish-fawn head, pale grey nape and pale brown upperparts. The upper belly is pale grey-fawn and lower belly and flanks pale chestnut. The tail is chestnut, the bill pale blue-grey. Legs are grey and the brown iris is surrounded by a grey eye-ring.

Where to see: Occurs on Bali and usually found in lowland open grassy scrub and rice fields, and often even in the centre of towns.

Java Sparrow *Padda oryzivora* 16cm

Indonesian name: Gelatik Jawa

Adults are largely grey, with a black head and throat, large white cheek patches, a black tail, white undertail-coverts, and a heavy pink or red bill. Juveniles are brownish-grey with a buff or brown breast. This once common bird has declined alarmingly in recent years.

Where to see: Endemic to Java and Bali. It is usually sedentary, found in small groups feeding on rice and seeding grasses, and in other low vegetation, mangroves and scrub.

Occasionally it forms large wandering flocks feeding on grain and rice. Sometimes seen close to human habitation, in cities, towns and villages.

Grey Wagtail *Motacilla cinerea* 17–20cm

Indonesian name: Kicuit Batu

A typical wagtail in shape and tail-wagging behaviour, running and skipping among the rocks and running water. Its flight is markedly undulating. It has a grey crown and mantle, with bright yellow underparts and rump. The chin and eye-stripe are whitish and the ear-coverts greyish. The underparts of juveniles are much paler. The very long, black tail is edged white. The bill is dark brown and the feet are pinkish-grey.

Where to see: Overwinters throughout the region. Prefers stony river beds, streams and damp meadows, from the coast to tops of mountains.

Eastern Yellow Wagtail *Motacilla tschutschensis* 17–18cm

Indonesian name: Kicuit Kerbau

Many variations in plumage occur amongst the subspecies, especially in winter plumage, making identification difficult. Of typical wagtail shape and behaviour, the species is olive-brown or olive-green on the back, with yellow underparts. The commonly occurring *M. t. simillima* has a grey crown, yellow throat and white supercilium, and any variations could relate to other, more rare subspecies. Immatures have browner upperparts and whiter underparts.

Where to see: Occurs as a winter visitor throughout the region. Commonly found in coastal lowlands, particularly pastures, rice fields and dried-up marshland.

Paddyfield Pipit *Anthus rufulus* 15–16cm

Indonesian name: Apung Sawah

Very similar in appearance to Richard's Pipit (*A. richardi*) but is generally paler and has shorter legs. This very ordinary looking pipit has upperparts streaked grey-brown with lightly streaked buffish underparts becoming white on the belly. It has a dark brown malar and submoustachial stripes and a long and broad buff supercilium that tapers off towards the back of the head. The tail is dark brown, as is the iris. The legs are flesh-coloured with a very long hind claw.

Where to see: Occurs throughout the region in open short grassland, rice fields and frequently by the roadside and close to cultivated land.

GLOSSARY OF TERMS

Canopy Unbroken layer of branches and foliage at tops of trees in forest

Cere Bare fleshy or waxy protuberance at base of upper mandible, including nostrils

Coverts Small feathers at base of quill feathers forming main flight surfaces of wing and tail

Dimorphic Occurring in two genetically determined plumage forms

Echolocation Navigation by ultrasound radar

Endemic Indigenous species restricted to a particular area

Eyebrow Contrasting line above eye (supercilium)

Eye-stripe Contrasting line through eye

Flank Side of the body

Frontal shield Skin or hard unfeathered area on forehead which extends to bill

Frugivorous Subsisting on a diet of mainly fruit and berries

Greater Sundas Borneo, Sumatra, Java and Bali, including offshore islands

Gregarious Frequently occurring in groups

Gular pouch Bare fleshy patch of skin around neck of cormorants and hornbills

Gunung Indonesian name for mountain

Hackles Long narrow and often pendulous feathers around neck

Lores Area of feathers between bill and eye

Mandible Upper or lower half of bill

Malar The area between base of bill and side of throat

Mesial Dividing down the middle

Migrant Non-resident traveller

Montane Relating to mountain habitats, usually above 900m

Narial feathers Small, bristly feathers projecting forward over the nostril, collectively called narials

Necklace A line of markings around front of neck

Orbital ring Unfeathered bare ring around eye

Primaries The main outer flight feathers (longest part of folded wing)

Primary forest Original natural forest

Race Another name for subspecies

Rackets Paddle-shaped ends to tail feathers

Resident Remaining in a local area throughout year

Roost Resting or sleeping place

Secondaries The inner flight feathers on rear half of wing

Secondary forest New-growth forest replacing primary forest

Spatulate Having thickened, rounded ends

Speculum Contrasting iridescent patch on a duck's wing

Sub-montane Relating to mountain habitats, usually below 900m

Subspecies A population morphologically different from other populations of same species

Sub-terminal Area above the end or tip of the tail

Supercilium A stripe above eye (eyebrow)

Terminal At the end or tip

Underparts Underside of body, from throat to undertail-coverts

Undertail-coverts Small feathers below tail covering bases of tail feathers

Underwing-coverts Small underwing feathers covering bases of primaries and secondaries

Upperparts Upper surface of body

Vent Area around anus, including undertail-coverts

Wattles Brightly coloured bare skin hanging from head or neck

Wing-bar A visible line of colour at tips of the wing-coverts

Wing-coverts Small feathers on wing covering bases of primaries and secondaries

BIBLIOGRAPHY

Useful Websites:

The Cornell Lab of Ornithology – A superb website to extend your interest in the enjoyment of birds, www.birds.cornell.edu/home/

Fat Birder – A good source of information on Birds in Indonesia, www.fatbirder.com/world-birding/asia/republic-of-indonesia/java/

Avibase – The World Bird Database, www.avibase.bsc-eoc.org/checklist.jsp?region=IDst

Xeno-canto – Sharing bird sounds from around the world – Bird Sounds of Indonesia, www.xeno-canto.org/set/1460

Burung Indonesia – The Indonesian partner of BirdLife International, www.birdlife.org/asia/partners/indonesia-burung-indonesia

Book References:

Eaton, J. A., van Balen, B., Brickle, N. W. & Rheindt, F. E. 2021. *Birds of the Indonesian Archipelago: Greater Sundas and Wallacea*. Lynx Edicions, Barcelona.

MacKinnon, J. & Phillipps, K. 1993. A *Field Guide to the Birds of Borneo, Sumatra, Java and Bali*. Oxford University Press, Oxford and New York.

del Hoyo, J. Elliott, A., Sargatal, J. & Christie, D. A. *Handbook of the Birds of the World*, 17 volumes. Lynx Edicions, Barcelona.

Checklists:

BIRDS OF SUMATRA
https://avibase.bsc-eoc.org/checklist.jsp?lang=EN&p2=1&list=ioc&synlang=&region=IDstsu&version=text&lifelist=&highlight=0 (viewed 01/09/22)

BIRDS OF JAVA
https://avibase.bsc-eoc.org/checklist.jsp?lang=EN&p2=1&list=ioc&synlang=&region=IDjact&version=text&lifelist=&highlight=0 (viewed 01/09/22)

BIRDS OF BALI
https://avibase.bsc-eoc.org/checklist.jsp?lang=EN&p2=1&list=ioc&synlang=&region=IDlsba01&version=text&lifelist=&highlight=0 (viewed 01/09/22)

PHOTO CREDITS

All photographs are copyright © the author, except for those listed below.

Bloomsbury Publishing would like to thank the following for providing photographs and permission to use copyright material.

Key: T = top; C = centre; B = bottom; L = left; R = right; BL = bottom left; BR = bottom right; TL = top left; TR = top right.

AA = Aurélien Audevard; APL = Agami Picture Library; AS = Amar-Singh HSS; CWM = Choy Wai Mun; CF = Con Foley; CG = Chris Galvin; DS = Dubi Shapiro; G = Getty Images; GM = Gilles Martin; JE = James Eaton; JO = János Oláh; KW = Karyne Wee; LP = Lars Peterson; LS = Laurens Steijn; M = Minden Pictures; MK = Matthew Kwan; MKG = Mohit Kumar Ghatak; MPW = Michelle and Peter Wong; NM = Neil Bowam; PE = Peter Ericsson; SS = Shutterstock.

Front cover TL CF, C CF, TR CF, B JE; **Back cover** T MK, C SS, B SS, B (large) SS; **1** MK; **3** SS; **10** SS; **12** B SS; **13** SS; **15** T LS/APL; **16** B MK; **17** T SS, B MK; **18** T SS, B MK; **19** CF; **20** B MPW; **21** T SS, B CF; **23** T AS; **24** T SS, B LP; **25** T SS, B SS; **26** T PE, B PE; **27** B AS; **28** B CG; **29** T SS; **31** MK; **33** B SS; **34** T AS, B SS; **35** SS; **36** T CWM, C CWM, B SS; **37** T MPW, B MPW; **38** T AS, B MK; **39** T MK, B SS; **40** T MK; **41** T SS; B DS; **42** B MK; **43** T CF, B SS; **44** T AS; **45** T CF, B MPW; **46** CF; **47** TL CWM, TR CWM, B CWM; **48** T SS; **49** T MK, B MK; **50** T MK, B MK; **51** TL MK, TR MK, B MK; **52** MK; **53** T AA/APL, B MPW; **54** T SS, B SS; **55** T SS, B MK; **56** T SS, B SS; **57** T SS, B MK; **58** T SS, B SS; **59** T MK, B SS; **60** T GM/G, B MK; **61** T MK, B MK; **62** T MK, B MK; **63** T SS; **64** MK; **65** T SS, B CWM; **66** TL MK, TR MK, BL MK, BR MK; **67** B MK; **68** MK; **69** B SS; **70** T CF, B MK; **71** SS; **72** MK; **73** MK; **74** MK; **75** T SS, B MK; **76** B SS; **77** T SS, B SS; **78** T CF, B MPW; **79** MK; **80** B MK; **81** TL CF, TR CF, B SS; **82** B SS; **83** TL MK, B MK; **85** TL SS, BL CWM; **86** SS; **87** T SS, B MK; **88** MK; **89** T SS; **90** SS; **91** B SS; **92** B SS; **93** T SS, B SS; **94** T SS, B SS; **95** B SS; **96** B MK; **97** T SS; **98** MK; **99** SS; **100** B SS; **101** B SS; **102** T CF, B SS; **103** AS; **104** T AS; **105** B AS; **106** B MPW; **107** T LP, B LP; **108** T SS; **109** B SS; **111** S; **112** SS; **113** T SS, B SS; **114** SS; **115** B SS; **116** T SS, B SS; **117** T SS, B SS; **118** T SS, B CF; **119** B MK; **121** B AS; **122** TL MK, TR MK, B MK; **124** T CWM, B SS; **125** MK; **126** T AS, B CF; **127** T CF, B CF; **128** T DS, B SS; **129** B MPW; **130** T JO, B SS; **132** T SS; **133** SS; **134** B SS; **135** T SS, B CF; **136** L CF, R CF; **137** B MPW; **138** T MPW, B SS; **139** SS; **140** B MK; **141** CF; **142** T MK, B SS; **143** T SS, B CF; **144** T SS; **145** T SS, B PE; **146** T KW, B CF; **147** T SS; **148** T MPW; **149** TL AS, TR AS, B AS; **150** T SS, B SS; **151** B AS; **152** SS; **153** B SS; **154** B SS; **155** T CWM, C MPW, B AS; **156** T NM/M, B AS; **157** T MK; **158** T MK; **160** T MK, B SS; **161** T SS, B SS; **162** T AS, B AS; **166** T SS, B SS; **167** B MK; **168** T MK, B CWM; **169** T SS; **170** B MPW; **171** B SS; **172** T SS, B SS; **173** B SS; **174** B SS; **175** T CWM, B CWM; **176** T SS, B JE; **177** B SS; **178** B SS; **179** T MK, B MK; **180** T SS, B DS; **181** T SS, B SS; **182** T SS, B SS; **184** T SS; **185** T SS; **186** SS; **187** T MK, B SS; **188** T MKG, B SS; **189** SS; **190** T SS, BL MPW, BR MPW; **191** T SS; **194** SS; **195** SS; **196** T AS, B MK; **198** T CWM, B CWM; **199** SS; **200** T SS, B SS; **201** B SS; **202** SS; **203** T PE, B MK; **204** MK; **205** T SS; **206** B SS; **207** B PE; **208** B SS; **209** T SS; **211** T SS; **212** B SS; **213** T SS; **214** T SS; **215** B SS; **216** B SS; **217** T SS, B MK.

INDEX